Can a Smart Person Believe in God?

Michael Guillen, PhD

NELSON BOOKS
A Division of Thomas Nelson Publishers
Since 1798

www.thomasnelson.com

Published in Nashville, Tennessee, by Thomas Nelson, Inc.

Library of Congress Cataloging-in-Publication Data

Guillen, Michael.
Can a smart person believe in God? / Michael Guillen.
  p. cm.
Includes bibliographical references.
ISBN 0-7852-6024-2
1. Faith and reason—Christianity. I. Title.
BT50.G85 2004
261.5'1—dc22   2004014212

Printed in the United States of America

04 05 06 07 08 QW 5 4 3

To my sisters Delores and Deborah,
who, like God, have always loved me,
flaws and all.

Contents

ONE

Depth Perception

The eye is the lamp of the body.
If your eyes are good, your whole body will be full of light.
But if your eyes are bad, your whole body will be full of darkness.

—Matthew 6:22–23

Some years ago, as ABC News's science editor, I participated in a TV special hosted by Peter Jennings on the subject of prejudice. In one of the segments, we did an experiment featuring a schoolteacher and large group of students.

The teacher began by dividing the kids up into two camps: Blue Eyes ("Bluies") and Brown Eyes. Then she proceeded to explain that Bluies tend to be slower, clumsier, and dumber than other kids.

To reinforce the lesson, every time a blue-eyed kid made the slightest mistake—much to the delight of all the Brown Eyes—the teacher said something disparaging like, "What else would you expect from a Bluie?" As you can imagine, after just a few minutes of this, most of the blue-eyed kids were thoroughly cowed, and some were even in tears.

I tell you this story because we, too, are split up into two camps: those who believe in God and those who believe in something else. You'll notice I resisted lapsing into the common practice of referring to the two camps as Believers and Nonbelievers; doing so would encourage the totally erroneous notion that "believing" or "having faith" is something only some of us do. Truth is, every one of us "believes." Every one of us "has faith." What divides us are the different objects of our faith, our different gods.

According to a *Newsweek* poll conducted in 2000 by Princeton Survey Research Associates, 94 percent of us—myself included—believe in the existence of God. (According to George Gallup Jr. and D. Michael Lindsay, in *Surveying the Religious Landscape: Trends in U.S. Beliefs*: "Over the past fifty years of research, the percentage of Americans who believe in God has never dropped below 90%.") This is an astonishing percentage, considering the notion of an unseen deity who created and possibly still interacts with the universe is (let's be honest) pretty far-out. You can't get that many people to agree on even relatively mundane issues, such as: "What is the exact color of this paper?" or "Is it better to unroll toilet tissue from the bottom or the top?"

According to the same poll, 4 percent of Americans don't believe in God—a pretty meager percentage, given the disproportionate attention and clout this camp appears to enjoy in today's secular age. The poll didn't ask what they do believe in, but from my experience it would certainly include Randomness, a god whose supernatural-like powers can allegedly transform complete chaos into exquisite order.

Each camp has powerful arguments to defend its respective deity, all too often using them to put one another down—as in the blue and brown eyes experiment. That's why I've written this book: first, to contribute some civility to the overall debate, and second, to rebut the argument that those who believe in God are dumber than those who do not. I hear this unseemly and unfounded prejudice voiced a lot these days, mainly from secular humanists who see themselves as smart, free-thinking realists and believers in God as dim-witted, superstitious sheep.

The accusation is expressed in many different ways, but its underlying message is always the same: you can't possibly be an enlightened, scientifically literate person alive in the twenty-first century and still believe in God, or in all the celestial trappings that go with Him. It's as if, as we venture forth into the new millennium, there's a gigantic Dante-esque sign overhead that reads: "Abandon Faith, All Ye Who Enter."

The Soviet cosmonaut Yuri Gagarin, first person to orbit the earth, quipped while in space: "I don't see any god up here." Similarly, a machinist from Toledo, Ohio, remarked: "No, I don't believe in God—[after all] did the space travelers ever see heaven in their trips?"

God is now obsolete, declares the modern-day Doubting Thomas, superceded by deities of a more earthly variety. The report of an MIT professor's imperious reaction to a campus event being sponsored by a Christian group says as much: "We don't care that these people are here," he reportedly scoffed. "At MIT, science and technology are the gods we serve."

Coming face-to-face with such bold atheism from a person with such impressive academic credentials can be intimidating, no matter who you are. It's the rare person among us whose confidence in God is so utterly rock-solid it can't be secretly shaken by some overzealous humanist accusing him of being a bonehead for having faith in something that allegedly doesn't jibe with today's scientific/intellectual paradigm.

I pray and fully expect that if you believe in God, then after reading my book, you'll feel more secure in the face of such confrontations. It contains many different lessons in how to defend and strengthen your faith—practical lessons I've learned from my lifelong association with *both* camps.

IQ Versus SQ

Who am I, and why should you care what I have to say? I'm the offspring of two very different worlds: one intensely intellectual, the other intensely religious. One a world of logic and IQ, the other of spirit and what I call *SQ.*

IQ, as everyone knows, stands for Intelligence Quotient. It's a measure of our ability to perceive the relatively obvious, physical aspects of reality; to solve problems and acquire conviction *intellectually*.

SQ stands for Spiritual Quotient, that ineffable, instinctive aptitude I believe we're all born with. I claim SQ is a measure of our ability to perceive the subtler, *non*physical aspects of reality; to solve problems and acquire conviction *spiritually*.

I've even created a quiz to test your Spiritual Quotient. It's not a rigorous examination, but it will give you a rough idea of how well developed you are in this regard.

And just in case you're wondering whether you should care, consider the recent explosion of scientific research on the subject. In 2002, sociologist Byron R. Johnson and his colleagues at the University of Pennsylvania's Center for Research on Religion and Urban Civil Society wrote an article titled "Objective Hope," in which they evaluate 498 studies published to date in peer-reviewed journals. The vast majority of the studies—anywhere from 65 to 97 percent of them—show that spirituality is associated with a long list of physical, mental and emotional benefits: lower levels of hypertension, depression, suicide, sexual promiscuity, drug addiction, alcohol abuse, and criminal delinquency, plus higher levels of academic achievement and feelings of well-being, hope, purpose, meaning, and self-esteem—as well as a longer life span.

I'll have much more to say about all of this later in the book, but for now, the message appears to be clear: your SQ should matter a great deal to you, the way, say, your cholesterol level, weight, and personality traits do—if not more so.

One word of caution. The words *spiritual* and *religious* have come to mean very different things to people. It's often hard to pin down those differences, but listen to this true story, told to me by my dear friend Chantal Westerman, the former ABC-TV entertainment editor who currently hosts television shows of a spiritual nature.

Among her many charitable extracurricular activities, Chantal spends time counseling high-risk violent offenders serving time for everything from carjacking to murder. During one of her visits, just as she was getting started, an inmate interrupted her, wanting to know: "Are you one of those religious types who's gonna preach to us?"

Chantal explained her visit had more of a *spiritual* nature, to which the inmate growled, "Religious, spiritual—what's the difference?" Before Chantal could respond, from the back of the room came the powerful voice of a large, tough-looking inmate: "I'll tell you the difference. Religion is for people who are afraid of going to hell. Spirituality is for those of us who've already been there."

For many, being religious has come to mean being a traditional churchgoer, an image that evokes both good and bad connotations—especially bad, given the ills that have befallen organized religion lately. That's why many today prefer the word *spiritual*; it comes without all that old baggage. For them, being spiritual connotes having a truly personal relationship with God, one that's unburdened by interdenominational dogmas and ceremonial formalities.

All that notwithstanding, I've decided to use the words *religious* and *spiritual* interchangeably. Why? Because for my purposes, the distinctions among people who believe in God are far less significant than the yawning difference between those who believe in God and those who don't.

You needn't agree with my decision—none of my arguments hinges on it. Just consider yourself alerted. Truth in advertising: in this book, I associate both spirituality *and* religiousness with aspects of reality that transcend what the mind alone can understand fully.

One other thing: I easily could write a book (and might one day) about the enormous value of logic and IQ. Analytical thinking has been, and continues to be, an essential part of my life, not to mention the intellectual development of our entire species. But alas, logic is largely useless when it comes to making sense out of the most significant events of everyday life.

Albert Einstein, who believed in God—though not the personal God of Judaism, Christianity, or Islam—expressed the crucial importance of SQ in trying to make sense out of our daily existence: "Strenuous labor and the contemplation of God's nature are the angels which, reconciling, fortifying, and yet mercilessly severe, will guide me through the tumult of life."

Pity the brilliant attorney with an underdeveloped SQ, for example, whose beautiful young wife is killed in a car accident, leaving him alone to raise their infant son and agonize over the seeming cruelty and capriciousness of life. Pity the brilliant but low-SQ biologist who spends her whole life studying the human retina yet has nothing to credit for its spectacular design except the supposedly fortuitous machinations of a mindless, purposeless universe. And pity the brilliant but low-SQ doctor who has no one and nothing to thank in the face of a spontaneous cure that defies all medical explanation.

Mind you, these folks don't have a clue they're missing anything important. Because of their SQ-challenged perceptions of reality, they aren't able to relate to God any more than fishes can to a warm fire. Rather, from their point of view, it's the rest of us who need the pity—the vast majority of the population that allegedly medicates itself with soothing religious delusions.

My Two Worlds

My life began in East Los Angeles, in the heart of the Mexican barrio. Dad was a Pentecostal minister, Mom a housewife.

Both of my parents were preachers' kids. Mom's father became a minister after seeing a group of church people lay hands on his son, my Uncle Carlos, reportedly healing him of crippling polio. Dad's father was called to the ministry after walking into a small church during an unscheduled stop in Kingsville, Texas, and being profoundly affected by its pastor's sermon. Grandpa then went on to become the legendary president of the CLADIC (*Concilio Latino Americano de Iglesias Cristianas*), the oldest independent Spanish-speaking Pentecostal organization in the United States.

I grew up attending services practically every day, or so it seemed, at *Templo Bethel*, the CLADIC church nearest to us. During that time I had the extreme good fortune of being surrounded by Christians with soaring SQs. Many of them were poor and uneducated, but they were (and still are) the intelligentsia of spirituality: by far the most humble, joyful, loving people I've ever met in my life.

Everyone around me assumed I'd follow in my father's and grandfather's footsteps and become a minister. But in the second grade, I was already nursing the unlikely dream of becoming a scientist and winning the Nobel Prize.

Where did this heady ambition come from? That's the $64 million question. At the time, I'd never met a scientist, never been encouraged to become one by a teacher or anyone else, and never been inside any sort of laboratory.

I have a theory, which I came to only recently and will share with you at the end of the book; but for most of my life my scientific aspirations were a complete mystery to me, as they were to my entire extended family. None of them had the slightest interest in science or, except for Dad, had ever even gone past high school.

Despite all that, I grew up to realize my vision, journeying from East Los Angeles to UCLA to Cornell and ultimately to Harvard. In the process, however, my IQ almost entirely crowded out my SQ. It's not that I became a registered, card-carrying atheist. Instead, I became what I've heard described as a *practical* atheist: someone who believes in God but lives as if he doesn't.

At any rate, I became a very different person. I stopped seeking the company of high-SQ people, stopped reading and thinking about the Bible, stopped discussing religion with friends, stopped putting God first in my life.

It was more comfortable for me that way. Most physicists don't believe in God (by the way, this is not because they know something the rest of the world doesn't, but because science— with its singular emphasis on IQ—attracts more than its fair share of people with underdeveloped SQs), so for the sake of fitting in, I became, like the Woody Allen character Zelig, a master of assimilation.

At parties or informal gatherings, I was always careful to hold my tongue whenever irreligious peers scoffed at the purportedly quaint notion of God and the allegedly emotionally crippled masses who believe in Him. I nodded my head compliantly, while they doubted a just and loving God could possibly exist in the face of so much undeserved and seemingly inexplicable human misery.

In 1984, mere weeks after I started teaching physics at Harvard, I received a phone call out of the blue (the way so many pivotal events in my life have happened). A producer at WCVB, Boston's ABC-TV affiliate, was casting around for a scientist to host a half-hour TV special and wanted to know if I was interested. Was I interested!?

The show, which earned huge ratings and an Emmy, eventually prompted WCVB to hire me part-time to do science reports for the evening news. Several years later, again out of the blue, I was invited to join the major leagues: ABC-TV hired me to become its national science correspondent. I continued to teach at Harvard, but suddenly the entire country became my classroom.

Regrettably, this lofty, new position served only to encourage my spiritual cowardice. My primary job as a reporter was to be objective, I told myself, so more than ever, both on air and behind the scenes, I studiously shied away from giving my personal take on any subject, especially religion.

But then it happened: one day, I went public with my belief in God. I'll never forget it. It was on February 25, 1997, during a roundtable discussion on *Good Morning America* concerning the scientific, legal, and moral implications of cloning. A few days earlier, Scottish scientist Ian Wilmut had stunned the world by announcing he'd cloned a sheep named Dolly.

As the roundtable discussion wound down, with only a few seconds left before going to commercial, moderator Charlie Gibson quickly turned to me and asked for my personal opinion. I told him the idea of cloning mammals, though exciting, troubled me, "not only as a scientist, but a scientist who happens to believe in God."

I worried that viewers would not react well to my religious confession. But the outpouring of calls and mail I received was more than supportive—it was positively giddy. In the popular imagination, you see, scientists are godless, so viewers had simply assumed I was an atheist and were absolutely thrilled to learn otherwise.

Emboldened by this unexpected public response, I put together a speech designed to encourage others to come out of the closet. I was eager for them to understand what it'd taken me a lifetime to discover, namely: believing in God is not anything to be ashamed of, nor in conflict with intelligence and modern-day education.

Now, years later, that speech has evolved into this book.

What I Believe

As a Christian, I believe in the monotheistic God of Abraham, Isaac, and Jacob—the God of the Book. The One who created the universe. The One who seeks a close, loving relationship with us, His most beloved creation. The One who first sent emissaries, then finally took on the form of the teacher Jesus of Nazareth, arguably the most extraordinary individual ever to walk the planet—all so our puny minds might begin to grasp His true, forgiving nature. The One who defines standards of behavior that invite us to transcend our baser, more destructive tendencies and experience not just fleeting, sensual pleasures but permanent, deeply satisfying happiness.

I recognize there are differences of opinion, some of them seemingly quite significant, even among us who believe in basically this same God. But the operative word here is *seemingly*. I have enormous confidence in my own particular beliefs about the God of Abraham; yet in the name of humility, I allow for the possibility that some of them are mistaken. That's why I honor any and every sincere variation on our shared impulse to worship Him.

For that matter, I respect the right of any person to hold *any* religious belief—even one I strongly oppose—so long as he practices it within the bounds of the law and human decency.

That includes atheism. In fact, the main reason for this book is not to rebut atheism (although, inevitably, I do that) but to discredit the arrogant manner in which its proponents often present and defend it—especially these days, when being cool often means coming across as sassy and self-reliant.

What You Believe

What about you? To which camp do you belong? For example, which do you find yourself relying on most, particularly in times of personal crises: IQ or SQ? Do you believe in God with all your mind and all your soul? Or do you believe in God with only your soul and find that in today's climate of "science think" and rampant cynicism your mind is barely managing to hang on?

Do you find it difficult or impossible to believe in anything you can't prove with IQ alone—and find the notion of SQ way too foreign and fuzzy? Or, put more bluntly, do you think believing in God makes about as much sense as believing in the Tooth Fairy?

Whoever you are, rest assured it isn't my purpose to convert you to my way of believing but to encourage you to recognize your capacity for IQ *and* SQ; to help you see the importance of developing, not one at the expense of the other, but both at once.

If you're someone in the first camp, whose members believe in God, I urge you:

- Please don't shun the importance of IQ and critical thinking in your relationship with God and His creation. Don't forget, we were made in His image, and that includes our brains.

- Please don't allow this scientific day and age to silence, undermine, or outright destroy your faith in God.

- Please don't let yourself be hoodwinked into thinking IQ is somehow superior to SQ.

- Please don't let science compare you to a computer. Thankfully, we aren't ruled solely by cold, calculating logic.

If you're someone in the second camp, whose members believe the universe is nothing more than a breathtaking accident, I urge you:

- Please don't overlook or underestimate the perceptive powers that come with a well-developed SQ. Seeing is a form of believing, but also, believing is a form of seeing.

- Please don't try fooling yourself or others about why you don't believe in God; whatever the reasons are, they have nothing to do with intelligence.

- Please stop hyping the successes of science; they are legion, but none of them has ever proved, or ever will be able to prove, that God doesn't exist. (Neither will science be able to prove that God *does* exist. Science as we know it will always be completely *neutral* on the subject.)

- Please reflect on your normal reaction to a gourmet feast at a five-star restaurant: you don't just rave about the meal or merely delight in analyzing its ingredients; at the end of the night, you acknowledge and praise the one who created it, even if you never actually see him. Answer this: why won't you allow yourself to do the same concerning this sublime universe of ours?

Seeing with Both Eyes

Whoever you are, I claim your journey through life works best when both your IQ and SQ are operating well, the same way your vision does when your two eyes are seeing well. In other words, there's more to reality than meets one eye.

Don't just take my word for it—try it. Right now, cover up one of your eyes and start walking around. See what I mean? Suddenly everything looks flat. You can't judge distances, because you've lost your depth perception.

If you reach for a cup, you're liable to come up short. If you approach a chair with the intention of sitting on it, you're liable to land on your bum. In all ways, with just one good eye, life is more uncertain and scary.

Similarly, depend on just IQ or SQ, and you'll always experience only a flat, uncertain, unsettling reality. It's only with IQ *and* SQ that you're able to see in stereo, able to recognize that the universe has depth. It's only with them both that you're able see yourself and others in their full, multidimensional, God-given glory.

That's it, my central message—and why, in this book, I speak about persons in terms of where they fall in a grid defined by IQ and SQ. I imagine the grid to look something like the example on the next page.

In the chapters that follow, I elaborate on each type of person. For now, please let me issue another important alert. In the grid, and throughout the book, I use "IQ" to refer not strictly to a person's intelligence, but to the *role* of intelligence or education in the person's relationship with God.

A woman with a PhD degree and an Einsteinlike IQ, for example, might come to believe, as many Amish do, that intellectualism is actually a stumbling block in anyone's attempted

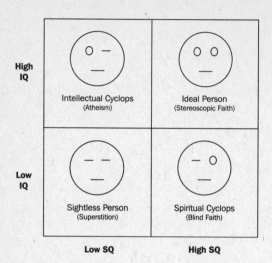

approach to God; that it's best to seek Him purely with faith. For that reason, such a woman, despite her enormous smarts, would fall into the low-IQ/high-SQ category of my grid. She'd qualify as a Spiritual Cyclops.

A man with only a high-school degree and an average IQ might believe, as I do, that the mind, like the spirit, is a gift from God and has a proper role to play in anyone's relationship with Him. For that reason, such a man, though not nearly as intelligent as the woman, would nevertheless fall into the higher-IQ category of my grid. He'd qualify as an Ideal Person.

This is the individual I believe we all should strive to become. The one who chooses to keep both eyes wide open, who practices *stereoscopic faith*.

I call this individual the Ideal Person because in Matthew

22:37–38 (and other places in the Bible), I believe God makes it very clear that in coming to Him we are to use both our SQ and IQ: "Jesus replied: '"Love the Lord your God with all your heart and with all your soul and with all your mind."This is the first and greatest commandment.'"

If you don't believe in the Bible, then please believe this: stereoscopic faith has changed my life profoundly—actually, profoundly doesn't even seem to cover it; it's more like *phenomenally* or *miraculously*—and is changing it still.

On the professional front, my stereoscopic faith recently spurred me to start my own production company, *Spectacular Science Productions*. It's a dream come true for various reasons, but mainly because it's increased my freedom to be creative and, above all, to go wherever the Lord calls me.

My stereoscopic faith is working gigantic changes in my personal life as well. After years of struggling with infertility, my wife, Laurel (whose own 3-D faith is as strong as mine, if not stronger), and I recently decided to adopt a young son. We're only four months into the adoption, but he's already brought us more joy than anything we ever imagined.

With both my IQ and SQ in full focus, I now feel light-footed, as though I'm skipping, not trudging, through life's uncertain journey. I no longer feel burdened by the endless worries that once kept me up at night and down by day; no longer at war with life, but at peace—the kind of inner, supernatural peace Saint Paul described as surpassing all understanding.

I'm not sure where my stereoscopic faith is leading me, but I'm thoroughly excited about the road it and God have me traveling. I'm more contented in my career than I can ever remember and happier in my relationships with family, friends, and coworkers. In general, I feel more alive, more enthusiastic, more optimistic about everything in life than I ever have before.

I wish the same good, true things for you. That's why I've written this book.

I pray that if you believe in God, you'll never again even *think* about being embarrassed to admit it to anyone, and that if you don't believe in God, you'll never again even *think* about looking down your nose at someone who does.

TWO

The Difference Is in the SQ

It is the stars as not known to science that I would know,
the stars which the lonely traveler knows.

—Henry David Thoreau

Can a smart person believe in God? The only reason the question even needs asking is this: persons identifying themselves as atheists, agnostics, humanists, and secularists—a total of less than 1 percent of the population, according to the City University of New York's 2001 American Religious Identification Survey—tend to see themselves as Marines of the mind: they are the few, the proud, the rational materialists.

Boasted the nineteenth-century American politician and atheist Robert Green Ingersoll:

"For ages, a deadly conflict has been waged between a few brave men and women of thought and genius upon the one side, and the great ignorant religious mass on the other."

In *Traveling Mercies: Some Thoughts on Faith*, best-selling author Anne Lamott describes how she became a born-again Christian.

It wasn't easy: it involved overcoming a sixties childhood domi-
nated by sex, drugs, alcohol, and snooty secularism:

> None of the adults in our circle believed [in God]. Believing
> meant that you were stupid. Ignorant people believed, uncouth
> people believed, and we were heavily couth. My dad was a
> writer, and my parents were intellectuals who went to the
> Newport Jazz festival every year for their vacation and listened
> to Monk and Mozart and the Modern Jazz Quartet. Everyone
> read all the time . . . We were raised to believe in books and
> music and nature.

Atheism and its attending superiority complex are especially
rampant among certain leading scientists. In an article titled,
"Scientists Are Still Keeping the Faith," published in the April 3,
1997 issue of *Nature*, Edward Larson and Larry Witham revealed
that about 40 percent of all American physical scientists believe
in a personal God (presumably, still more of them believe in a
nonpersonal God). Considering science's widespread reputation
for being godless, that's a pretty sizable fraction. But in a subse-
quent study, the authors discovered that among members of the
National Academy of Sciences—science's high priests—a mere
7 percent believe in a personal God.

During my years at Harvard, I recall a physics professor teaching undergraduates about the seminal contributions of the early
twentieth-century Cal-Tech physicist and Nobel Prize–winner
Robert Millikan. Millikan is renowned for his brilliant and historic oil-drop experiment, in which he discovered that every

electron carries an indivisible electric charge. It's too bad, lamented the Harvard professor, that Millikan, a devoutly religious man, was a such a "low-brow" (his exact words) when it came to his personal beliefs.

If you believe in God *and* the importance of intelligence and education *and* civility, how do you respond properly to such haughty atheism? For starters, by recognizing that for all their superior airs, atheists are really no different from you and me. Like us, atheists believe in something they can't prove scientifically.

Atheists Are Believers

The British actor and writer Quentin Crisp tells a funny story about the time he visited Northern Ireland and announced he was an atheist. Crisp recalls: "A woman in the audience stood up and said, 'Yes, but is it the God of the Catholics or the God of the Protestants in whom you don't believe?'"

The point is, it's impossible for an atheist to disbelieve in God without believing in some alternative—either that, or he must confess he believes in *nothing*. Even agnostics, who are somehow able to make it through life withholding judgment on one of life's most defining issues, have to believe in *something* during the interim. For most of them, atheism seems to be that something.

In my experience, atheists tend to believe in the cosmic existence of highly fortuitous accidents created by Randomness, which I write with a capital *R* because, for all intents and purposes, Randomness is the atheist's god. As the nine-teenth-century English poet Francis Thompson affirmed, "An atheist is a man who believes himself an accident." But within that broad definition of atheism, there are variations—denominations, if you will.

First off, there are the agnostics, whom I call Uncertain Atheists because by allowing for the possibility that God does exist, they admit they're not quite sure. For them, the jury is still out.

Next are the Arrogant Atheists, whom you heard from at the very beginning of this chapter. These are the low-SQ persons who worship Intellectualism—a supernatural faith, I might add, considering that our IQ's historical track record is, to put it politely, decidedly mixed. Above all, Arrogant Atheists feel a need to believe they're smarter than everyone else. (See chapters 6 and 7.)

Then there are what I call the Humble Atheists, persons who worship Intellectualism but are honest enough to admit theirs is not some superior belief. The prolific science-fiction writer Isaac Asimov once explained his Humble Atheism this way: "I've been an atheist for years and years, but somehow I felt it was intellectually unrespectable to say that one is an atheist, because it assumed knowledge that one didn't have . . . I don't have the evidence to prove that God doesn't exist, but I strongly suspect that he doesn't."

One denomination that particularly intrigues me consists of what I call Christian Atheists, persons who embrace my religion's values but not its God. They are those whom Saint Paul appeared to be speaking about when he predicted in 2 Timothy 3:1, 5: "But mark this: There will be . . . [people] having a form of godliness but denying its power."

One of the most prominent Christian Atheists I know is someone I respect highly. He is Edward O. Wilson, Harvard sociobiologist, Pulitzer Prize–winning author, and quite honestly one of the nicest, most decent human beings whom I have the honor of calling friend.

Ed was a born-again Southern Baptist until the age of seventeen or eighteen, when he began to lose his faith. He no longer believes in a personal God but says, "I have not abandoned my Christian upbringing. I'm very much a Christian in ideals and ethics, especially in terms of belief in fairness, a deep set obligation to others, and the virtues of charity, tolerance and generosity that we associate with traditional Christian teaching."

We must also not forget the Rebel Atheists, persons who reject God as part of an overall aversion to authority. Their god is Individualism. Take, for example, the American feminist Voltairine de Cleyre. After many failed attempts at suicide, de Cleyre died from illnesses brought on by a life of poverty—but not before she was able to boast: "I die, as I have lived—a free spirit, an Anarchist, owing no allegiance to rulers, heavenly or earthly."

Finally, I wish to recognize what I call the Atheist's Atheist, that rare person who worships neither God, Intellectualism, nor Individualism, but rather some other all-knowing, all-powerful abstraction. My favorite example of this is the nineteenth-century German philosopher Arthur Schopenhauer. In *A History of God*, author Karen Armstrong explains that for Schopenhauer there was "no Absolute, no Reason, no God, no Spirit at work in the world: nothing but the brute instinctive will to live."

I could go on naming atheism's other denominations, but I trust I've made my point. It really does appear impossible for anyone to go through life believing only in things he can prove scientifically or argue conclusively academically. One way or another, we are *all* believers in something that lies beyond the grasp of pure smarts.

Smart Believers in God

Contrary to the slanders voiced by Arrogant Atheists, we who believe in God are in very distinguished company indeed. The simplest evidence of this: according to a 2003 Harris Poll, among Americans with post-graduate degrees—in other words, our country's most well-educated men and women—a whopping 85 percent believe in God.

Surprised at the high percentage? You shouldn't be. At widely separate times and places, free-thinking, urbane, brilliant persons from Solomon and Leonardo da Vinci to William Shakespeare and Henry David Thoreau—with IQs exceeding those of most Arrogant Atheists—have all pondered the evidence nature and life have to offer and come to the same conclusion: there's more to reality than just space and time, matter and energy, logic and reason.

Even Albert Einstein, the quintessence of human intelligence, made that discovery. He didn't believe in God as any one particular religion defines Him and repeatedly declared he'd sold his body and soul to science. Yet he made it clear in a quote documented in Max Jammer's authoritative book, *Einstein and Religion*: "I'm not an atheist, and I don't think I can call myself a pantheist."

Einstein believed in one God and read the Bible often. Far from ridiculing our humble, religious inclination to stand in awe of the Designer of our universe, he defended it in this eloquent way: "In every true searcher of Nature, there is a kind of religious reverence, for he finds it impossible to imagine that he is the first to have thought out the exceedingly delicate threads that connect his perceptions."

Indeed, high-IQ/high-SQ people—individuals with stereoscopic faith—are the bona fide heroes and heroines of human history. In the fifth century, while barbarians ransacked the once-mighty Roman Empire, Catholic bishops, abbots, monks, and other clerics became Western civilization's chief protectors. They converted monasteries into schools, founded libraries, and manned copying chambers: places where they laboriously and lovingly duplicated by hand books both sacred and secular. Centuries later, their superhuman efforts, together with those of their Jewish and Islamic counterparts living in the Arab world, helped usher in the European Renaissance.

People with stereoscopic faith also founded the world's first modern colleges, among them the University of Paris, with its legendary Sorbonne, and the University of Oxford, the oldest institution of higher learning in the English-speaking world and home of the celebrated Rhodes scholarships.

As for the New World, one of the first things Puritans did when they arrived was set up Harvard College, the oldest university in North America. Afterwards, Christians of every denomination—from Roman Catholic, Baptist, and Congregationalist to Episcopalian, Methodist, and the Church of Christ—established what are today many of our most prestigious academic establishments.

High-IQ/high-SQ people and institutions striving to express their joyous belief in the supernatural have authored, developed, encouraged, cherished, perfected, and safeguarded countless innovations in art, music, literature, architecture, education, even science—all the finer things in life we call *culture*.

Christianity, for one, has been a chief patron of the most gifted musical geniuses of all time—from Bach and Beethoven to Mozart and Messiaen. Many popular musical forms, including the oratorio anthem, carol, hymn, ordinary mass, and requiem mass, have religious origins. Moreover, western choirs were invented in the sixth century in ecclesiastical song schools started by Pope Gregory I. Similarly, high-IQ/high-SQ architects are responsible for some of the most famous structures ever built, including five of the Seven Wonders of the Ancient World—among them, the great pyramids of Egypt, Temple of Artemis, and Mausoleum at Halicarnassus. There are also the Taj Mahal, Cathedral of Notre Dame, Westminster Abbey, Alhambra, and on and on and on. Entire architectural styles—Byzantine, Romanesque, Gothic—are a direct result of our irresistible urge to build beautiful, soaring, inspiring religious sanctuaries.

The dome, too, one of the most ubiquitous and beautiful of all architectural inventions, was created in response to our religious impulses. Representing the heavenly vault, a pure dome first appeared circa AD 120 atop the Pantheon of Rome, Emperor Hadrian's shrine to the gods of his people. And fourteen centuries later, at the behest of Pope Julius II, Michelangelo designed an exquisite, ribbed variation of it for Saint Peter's Basilica, a style of dome that today crowns capitol buildings the world over, including our very own in Washington, D.C., and states across the nation.

Finally, people with stereoscopic faith have played a seminal role even in the development of religion's supposed archrival, science. That's right. We hear a lot about the relatively few times religion and science have gone for each other's throats— for example, the infamous feuds over heliocentrism versus geocentrism and creationism versus evolutionism—but they are the exceptions. Fact is, without the long line of intellectual revolutions led over the centuries by men and women of high IQ and high SQ, science would not be what it is today.

Saint Albertus Magnus (patron saint of natural scientists and AKA Albert the Great), grandfather of modern geology; René Descartes, father of modern Western philosophy; Gottfried Wilhelm Leibnitz and Isaac Newton, codiscoverers of the calculus; Robert Boyle, cofounder of modern chemistry and the London-based Royal Society, one of the world's most elite body of professional scientists; Gregor Johann Mendel, father of modern genetics; Nicolaus Copernicus, father of modern astronomy, Georges Lemaître, father of modern cosmology;

Galileo Galilei, cofather (with Newton) of the Scientific Revolution: these were high-IQ/high-SQ people (many of them out-and-out *religious clerics*) who saw science as a way of studying God's creation. Today, their names constitute the Who's Who of Western science.

Low-SQ Religiosity

Despite their superior airs, in other words, atheists can't honestly say they're any smarter than people who believe in God; nor can they themselves be accused of being any dumber. Frankly, both camps consist of people whose IQs range across the intellectual spectrum, from really smart to really dumb. The real difference between them is their SQs.

I claim atheists have relatively low SQs because they worship merely the obvious: the human mind, nature, or the laws of science. As the famous American architect Frank Lloyd Wright once crowed: "I believe in God, only I spell it Nature."

By contrast, I maintain that people who believe in God have higher SQs because they're able to see deeper into reality, perceiving the *Creator* of that mind, that nature, and those laws. Mortimer Adler, one of America's foremost philosophers, put it beautifully: "Only members of the human species have the conceptual powers that enable them to deal with the unperceived, the imperceptible, and the unimaginable."

Still, by virtue of what SQ they do have—of their having gods, however superficial they might be—atheists *are* religious. During the nineteenth and twentieth centuries the

English journalist and Uncertain Atheist John Morley evidenced this fact by calling for science to step up to the plate: "The next great task of science is to create a religion for mankind."

Today, the famous science-fiction writer and atheist Kurt Vonnegut Jr. reveals a similar spiritual yearning: "I hope that many earthling children will respond to the first human footprint on the moon as a sacred thing. We need sacred things."

The first time I personally witnessed low-SQ religiosity in action was when, as a graduate student at Cornell, I heard the Uncertain Atheist astronomer Carl Sagan deliver a public lecture on the latest discoveries in astronomy. In typical fashion, Carl dazzled the standing-room-only crowd with his brilliant command of the subject and poetic eloquence. But afterwards, during the Q & A session, several attendees ridiculed him for using the words "majesty and awe" to describe the cosmos. How could he possibly have used such meaningless terms in a scientific lecture? they scolded.

Meaningless? Scientifically, maybe, but not religiously. I came to know (and respect) Carl well enough to discover that he had a deep, SQ-type reverence for the beautiful way in which our universe operates. True, he never looked far enough past that superficial beauty to catch a glimpse of its Author, but nevertheless he was, in his own way, a devoutly religious man.

The same kind of religiosity is even more evident in Carl's widow, Ann Druyan. Like him, she doesn't believe in God yet is spiritually passionate about the scientific process that elucidates the natural world. Listen to her in a speech describing an astronomy show she cocreated for the Hayden Planetarium in New York. She sounds positively evangelistic:

> This is what got me thinking about how we might offer something that would be at least as compelling as whatever anyone else in the religion business is offering. We get to take you through the universe . . . and teach you something about the nature of life. It's a very uncompromising message about evolution and I think very directly promotes the kind of values and ideas I think we share. Every kid who goes to a city public school gets taken to these shows (*Skeptical Inquirer*, November/December 2003, 25-30.)

And that's just for starters. Elsewhere in the speech, Druyan sounds like an out-and-out religious crusader: "Why don't we take over the planetaria of the country, of which there are hundreds and turn them into places of worship? . . . Why don't we take over these places and have services in the planetaria?"

Far from being exceptions to the religious rule, far from being coolheaded Marines of the mind, deep down inside, even atheists can't help but respond spiritually to what they comprehend only academically. It's as if they have just enough SQ to get an inkling of what persons with a higher SQ are able to perceive with greater clarity. Atheists don't have the SQ necessary to see past the obvious, but they do have enough to feel the tug that brings us all to our knees whenever we contemplate the "awe and majesty" of our extraordinary universe.

THREE

What's the Problem?

I don't believe in God, because I don't believe in Mother Goose.

—Clarence Darrow

To believe in God is impossible—not to believe in Him is absurd.

—Voltaire

In this chapter, I rebut some of atheism's most common complaints about theism. But before I do, I have a slightly embarrassing confession to make. Back when I was less mature spiritually and intellectually, I was actually afraid to read any of the atheists' arguments—afraid they might contain some revelation that would scuttle my faith in God, or at least cripple it.

One day, however, I came across a magazine published by secular humanists and mustered the courage to read it. That's when something strange happened. Much to my surprise—and please understand, I mean no disrespect, I'm just being honest—I found nearly all of their disputations to be naïve, even childish.

Too many of their authors, it seemed to me, even ones

with sophisticated educations whose arguments were filled with impressive-sounding technical jargon, had not done their homework, not even something as basic as studying the Bible. I had the impression many were merely parroting arguments they'd heard earlier but hadn't thought through for themselves.

After that, I went on to discover something else that surprised me. Many (maybe even most) atheists seem to have turned their backs on God for all the wrong reasons. I've had former atheists confess to me they were closet believers in God all along but were too rebellious, resentful, arrogant, or deeply scarred by negative childhood experiences to admit it. They were disgusted with organized religion, hate-mongering zealots, hypocritical churchgoers, bigoted or sexist religious leaders, child-molesting clergy, unscrupulous televangelists, and on and on.

Nowadays I still make it my business to keep up with the atheists' latest contentions, even though I must admit I get less and less out of doing so, both intellectually and spiritually. Some of their arguments are quite clever and even entertaining—I'll give them that. But most of them are just old arguments dressed up to look new.

One old chestnut, for instance, invokes our well-known human tendency to see patterns where none exist—as evidenced, for example, by our reading shapes into amorphous cloud formations or splotches of ink. Citing this, many atheists jump to the conclusion that people who notice evidence for God everywhere in nature are seeing something that really isn't there.

The problem with the argument is that loads of remarkable patterns do exist in nature—they're not illusory. Indeed, the entire edifice of science is based precisely on the existence of such bona fide patterns: we call them scientific laws. Atheists opt to believe all these patterns were authored by Randomness, but in doing so they're taking a personal leap of faith no more (or less) intellectually credible or defensible than the one taken by those who believe the patterns were authored by God.

Bottom line: I've yet to come across any atheist argument that injures my faith even a little bit. My faith is now far too strong and well informed for sophomoric syllogisms to threaten it. If anything, the overall feebleness of atheism's reasoning has served only to *strengthen* my belief in God.

If you already believe in God, I hope the following discussion will help strengthen your faith too. If you're an atheist, I hope you feel I've done justice to your arguments and that my rebuttals will challenge your mind and heart in some positive way.

Just Another Superstition?

In his excellent book *How We Believe: Science, Skepticism, and the Search for God*, the thoughtful Uncertain Atheist Michael Shermer reflects on how we humans always have been superstitious. For instance, during the Middle Ages: "If a noblewoman died, her servants ran around the house emptying all containers of water so her soul would not drown."

One of Shermer's more hilarious contemporary examples

involves former baseball player Wade Boggs: "[He always insisted] on running his wind sprints at precisely 7:17 P.M., ending his grounder drill by stepping on the bases in backward order, never stepping on the foul line when taking the field but always stepping on it returning to the dugout, and eating chicken before every game."

For most atheists, believing in God is as silly and superstitious as Wade Boggs's daily ritual. Boasted one atheist right after biologists announced their successful sequencing of the human genome (our genetic recipe): "Scientific discoveries puncture the clouds of superstition that surround human existence and weaken the grip of religion over the minds of men and women."

There's no denying it, we—and I do mean *all* of us—are superstitious. For example, when we sneeze, it's common for people nearby to say, "Bless you!" Ever wonder why? It's because during the Middle Ages, people worried the devil could sneak into them during that unguarded moment when their mouths were open and their eyes shut tight. Their only sure protection, they believed, was for people around the sneezers to invoke the name of God.

Even scientists are not above being superstitious. During the seventeenth century, Johannes Kepler and many other distinguished astronomers of the day practiced astrology right alongside astronomy. Isaac Newton and many of his distinguished peers practiced alchemy right alongside chemistry. Indeed, modern astronomy and chemistry grew out of those superstitious beliefs in astrology and alchemy!

But are atheists right—is believing in God itself a superstition? Is it really no different from believing coins tossed into a wishing well will bring us good luck or salt tossed over our shoulders will ward off bad luck? Are God and the devil merely the mothers of all lucky charms, good and bad?

Clearly, atheists who believe so are not familiar with human history; otherwise they'd know that time and again, religion has been the civilized world's principal agent against superstition, shepherding entire nations from gullibility and ignorance to pre-Christian and Christian sophistication.

Clearly, these atheists are also unfamiliar with the Bible; otherwise they'd know the God of the Book Himself is strongly opposed to superstition. For example:

- Leviticus 19:26: "Do not practice divination or sorcery."

- Deuteronomy 18:10–12: "Let no one be found among you who sacrifices his son or daughter in the fire, who practices divination or sorcery, interprets omens, engages in witchcraft, or casts spells, or who is a medium or spiritist or who consults the dead. Anyone who does these things is detestable to the LORD."

- 2 Chronicles 33:5–6: "He built altars to all the starry hosts [practiced astrology]. He sacrificed his sons in the fire in the Valley of Ben Hinnom, practiced sorcery, divination and witchcraft, and consulted mediums and spiritists. He did much evil in the eyes of the LORD."

- Isaiah 2:6, 8: "You have abandoned your people, the house of Jacob. They are full of superstitions from the East; they practice divination like the Philistines and clasp hands with pagans . . . Their land is full of idols; they bow down to the work of their hands."

According to a Spring 2001 Gallup poll, more than a quarter of today's Americans believe in witches and astrology, and more than a third in ghosts. But please don't lump those beliefs in with our belief in God, as some atheists do. As we've just seen, to God, superstitions of any kind are a grave sin, an affront to everything He stands for and wishes us to be.

The Roman philosopher Seneca put it this way: "Religion worships God, while superstition profanes that worship." I put it this way: Belief in God is the antithesis of superstition. At its best, it represents supremely high IQ and high SQ, whereas superstition represents low IQ and low SQ. The two can't possibly be more different.

Contagious Disease?

Our belief in God is nothing if not incredibly stubborn; it perseveres even in today's highly cynical, highly intellectual climate. Atheists have a hard time explaining this, yet they keep trying.

In his book *A Devil's Chaplain*, Oxford University evolutionary biologist and unabashed Arrogant Atheist Richard Dawkins claims humanity's belief in God persists simply because generations of religious parents keep infecting the impressible minds of their offspring with a stealthy virus. Yes, that's right, *a virus*.

This imagined God pathogen is so deeply ingrained in our brains, Dawkins maintains, we don't even realize we're infected: "The patient typically finds himself impelled by some deep, inner conviction that something is true, or right, or virtuous; a conviction that doesn't seem to owe anything to evidence or reason . . . We doctors refer to such a belief as 'faith.'"

Clever though it is, Dawkins's theory is not considered very credible, even among some atheists. In *How We Believe*, Uncertain Atheist Michael Shermer faults the theory's unseemly hostility toward God. Says Shermer: Dawkins unjustly portrays religion as "a *disease*, a *scourge* on humanity, which . . . we must rid ourselves of before it does us in."

Moreover, cognitive psychologist James Polichak eloquently spells out the theory's many scientific and logical flaws in his 1998 article in *Skeptic* magazine, "Memes—What Are They Good For?" In the technical literature, contagious viruses of the sort we're discussing are called *memes*.

I myself have at least three problems with the contagion theory. First, it's true most of us parents do place into the heads of our young, suggestible charges certain charming fantasies, for instance, Santa Claus, the Easter Bunny, and the Tooth Fairy. But somewhere on their way to adulthood, most kids soon discover the truth, either by accident or because the inconsistencies associated with keeping those fantasies alive reach a critical mass whereby they can no longer be believed.

What Dawkins's theory doesn't answer with any persuasion is this: if God is just another viral fiction drummed into young, gullible heads, why does He alone escape being disinfected? Why, especially in the face of so much modern skepticism, doesn't our belief in God go the way of our belief in Santa Claus, the Easter Bunny, or the Tooth Fairy?

Second, it certainly does appear as if most of us merely perpetuate our parents' beliefs—an impression Dawkins cites to defend his notion that religion is a communicable disease. But rather than a germ, other explanations such as straightforward inertia, familiarity, or habituation can easily account for this; we tend to do lots of things our parents did, including eating the same foods. So what?

Furthermore, a closer look at our religious proclivities reveals a picture exceedingly more complex than Dawkins paints. In *The Next American Spirituality: Finding God in the Twenty-First Century*, authors George Gallup Jr. and Timothy Jones explain: "Seven in ten Americans say their faith has changed significantly, with equal proportions saying it came about as a result of a lot of thought and discussion, and as a result of a strong emotional experience."

With so many people taking such a thoughtful, deliberate approach to God, the authors say, religion in the U.S. is becoming increasingly pluralistic. For example: "One writer described her early faith like this: Mine was a patchwork God, sewn together from bits of rag and ribbon, Eastern and Western, pagan and Hebrew, everything but the kitchen sink and Jesus thrown in." You might disapprove of this rising tide of so-called *pastiche spirituality*, but it does rebut Dawkins's assertion that we mindlessly replicate our parents' belief systems. Clearly we do not.

My third and final problem with Dawkins's theory is this: it fails to explain how the putative God virus is able to survive truly hostile climates, such as the one in modern-day China. For several decades, up until the 1980s, Communist party officials tried nearly everything in their power to destroy religion, to thoroughly indoctrinate Chinese kids into thinking that religion and God were nothing but backward, contemptible, and dangerous superstitions. "Triumph over ignorance with science!" was their battle cry.

If Dawkins were correct—that religion of any kind is a contagious illness passed on from one generation to another—then by now we'd expect the Atheism bug to have thoroughly infected the whole of China. But just the opposite is happening: atheism in the country is weakening and spirituality is strengthening.

That's right. After decades of Communist Chinese efforts to shame, suppress, and outright kill it, traditional religion is on the rise. Moreover, the revival seems to involve not just China's indigenous religions Buddhism and Daoism, but Christianity as well. "Christianity is today more vigorous than at the height of Jesuit influence in the seventeenth century," reports Arthur Waldron, Professor of International Relations at the University of Pennsylvania, in *Orbis*, the journal of the Foreign Policy Research Institute, "or at the peak of Protestant evangelism in the 1920s."

In the face of this swelling religiousness, report James and Evelyn Whitehead, Research Fellows at the University of San Francisco's Ricci Institute of Chinese-Western Cultural History, China's Religious Affairs Bureau recently caved in and finally allowed the publication of the Catholic church's one-thousand-page encyclopedia, censoring only seven entries, including ones on Atheism, Communism, and Socialism. In other words, God is not just surviving in a society bent on exterminating Him, He's actually flourishing!

Above all—please take notice, Mr. Dawkins—the Whiteheads report that increasingly even Communist intellectuals are no longer accusing religion of being a dangerous bourgeois infection. Instead, in universities all over China, the Whiteheads say, there's an explosion of new interest in God. Scholarly journals dedicated to the respectful study of religion are multiplying. In one of them, the Whiteheads report, a recent lead article titled "The Concept of God in an Age of Rationalism" contained a long, positive-minded quote from Pope John Paul II.

In my opinion, the simplest explanation for this phenomenon (and similar evidence from other communist countries) is unmistakable: Even where He's not officially welcome, God never goes away because the ubiquitous evidence for His existence never does. God exists, and no one—not even atheist political zealots using Draconian tactics to foist their godlessness on others—can keep individuals from recognizing that powerful truth, anymore than hardened soil can keep a seedling from pushing upward in its determined search for sunlight.

God's reality—not some hypothetical virus—is the most clear-cut explanation for why, long after we've abandoned our childish beliefs in Santa Claus, the Easter Bunny, and the Tooth

Fairy, we keep on believing in Him. Why, despite all the efforts by self-important atheists——who themselves merely come and go——He endures.

Comforting Delusion?

According to evolutionary biologists, religion is easily explained away with this theory: belief in God is but the neural equivalent of a Valium tablet, a protective illusion our brains invented to help us cope with the stresses of everyday existence. It's all about survival of the fittest: Without this comforting fiction, these scientists contend, we'd have less protection from life's madness, therefore less chance of surviving, therefore less chance of passing on our genes to the next generation. *Ta-da!*

We supposedly believe in God because we're too afraid *not* to believe, too afraid to face up to the possibility we're nothing but insignificant accumulations of star dust inhabiting a cold, purposeless universe. We supposedly believe in God to delude ourselves into thinking: 1) we're significant, 2) life has meaning, and 3) everything's going to be A-okay. We're like kids, atheists claim, and God is like our make-believe buddy——a comforting figment of our frightened imaginations.

Philosopher and atheist Paul Kurtz, founder of both the Council for Secular Humanism and the Committee for the Scientific Investigation of Claims of the Paranormal, puts it this way:

"Religious systems of belief, thought, emotion, and attitude are products of the creative human imagination. They traffic in fantasy and fiction, taking the promises of long-forgotten historical figures and endowing them with eternal cosmic significance."

Never mind that historical figures such as Moses, Jesus, Mohammed, and Buddha are anything but long forgotten; what about Kurtz's main charge? Is something seriously wrong with 90-plus percent of the American population? Are we all dupes for believing in God—poor, self-deluded weaklings who don't have the moxie to deal with the allegedly dismal truth of our allegedly pointless existence?

In the mind of George H. Smith, author of *Atheism: The Case Against God*, the answer is a resounding yes. Says he: "If the choice must be made between the comfort of religion and the truth of atheism, many people will sacrifice the latter without hesitation."

But wait a minute. *Comfort of religion?* Herein lies the main flaw in the atheists' "comforting delusion" argument. Were God some velvet teddy bear, their argument might hold water. But anyone who worships Him knows that He isn't.

Oh sure, God comforts us, especially in our times of need, but He also needles us, reproves us, and at times scares us silly. Just ask Moses, Job, or Jesus Himself. At critical times in their lives, these individuals found themselves feeling anything but comforted on account of their devotion to God.

Comfort of religion? Tell that to Peter, who reportedly was crucified upside down on account of this alleged neural protection mechanism. *Comfort of religion?* Tell that to Mother Teresa, who combed the gutters of the most squalid places on earth for patients dying from every affliction in the book, all on account of her devotion to this alleged figment of the human imagination. *Comfort of religion?* Tell that to the large number of Christians worldwide who are martyred annually on account of their unswerving allegiance to this alleged "advantage" bequeathed to them by evolutionary biology.

According to the *World Christian Encyclopedia*, edited by David B. Barrett and Todd M. Johnson, and published by Oxford University Press, roughly 165,000 Christians worldwide are martyred every year. In *World Christian Trends AD 30–AD 2200*, Barrett and Johnson explain that by "martyrs" they mean: believers in Christ who lose their lives prematurely because of their faith, in situations of witness, as a result of human hostility.

According to the International Religious Freedom Act, enacted in 1998 by the U.S. Congress, an even greater number of people—"[m]ore than one-half of the world's population"—suffer non-fatally for their faith. The persecutions include everything from "state-sponsored slander campaigns, confiscations of property, [and] surveillance by security police" to "torture, beatings, forced marriage, rape, imprisonment, enslavement, [and] mass resettlement."

Comfort of Religion? In the face of so much discrediting evidence, I find it exceedingly more plausible to believe that God actually exists than to believe in the atheists' fantasy that He is but a merciful fabrication—a neurochemical pacifier—bequeathed to us by random evolutionary forces for the purpose of making our lives easier and more palatable.

Frankly, I sometimes wish He *were* merely a figment of my imagination. Then I could order Him around, the way my

CAN A SMART PERSON BELIEVE IN GOD?

four-year-old son does *his* imaginary pals. As it is, the deity inside my head, and also my heart and soul, is a God of hugs and hammers. A God who in many ways makes my life harder, not easier. A God who always has the last word.

Comfort of religion? Yes, there's comfort in my religion. But as my former pastor, the Reverend Kenneth Valardi, once put it, "If our minds were truly in the business of creating God, there's no way they'd invent Christianity!"

A Deplorable Source of Evil?

Atheists' allegations that religion is evil and bloodthirsty have been repeated so often, and with such conviction, most of us don't even question it anymore. Here are some notable examples:

> Christianity . . . has brought an infinity of anguish to innumerable souls on this earth. (Joseph Conrad)

> I regard [religion] as a disease born of fear and as a source of untold misery to the human race. (Bertrand Russell)

> God is a mean-spirited, pugnacious bully bent on revenge against His children for failing to live up to his impossible standards. (Walt Whitman)

> If people can kill for God in this way, this is the best reason never to believe in God! (Anonymous, in wake of 9/11)

There's absolutely no denying we have a woeful history of committing unspeakable acts of violence in the name of our religious beliefs. Atrocities range from pagans sacrificing their children, to Jews persecuting early Christians, to Christians subsequently persecuting Jews, to radical Islamic fundamentalists attacking innocent Americans.

However, I think it's only fair to point out that throughout history, peace-loving men and women of every faith, including atheism, have sometimes found it necessary to resist evil with aggression because all other realistic options have been exhausted. A recent example of that, I believe, is America's retaliatory strike against the Taliban in Afghanistan.

Unfortunately, and often tragically, where resisting a clear-cut evil was *not* the main motivating factor, the true reasons for the violence we've committed in the name of religion, I claim, have been fear, ignorance, hatred, greed, jealousy, or any of the other myriad emotions God clearly condemns. He calls us to separate right from wrong—and remember, "Thou shalt not murder" is right up there in the Top Ten. Yet our baser instincts and rebellious natures constantly sabotage our ability to behave accordingly.

Indeed, our belief in God, our SQ, is ultimately not any more to blame for our violence than our intelligence. After all, it took a whole lot of IQ for the 9/11 terrorists to plan and execute their devilishly clever attack.

Nazism, too, had more to do with IQ than SQ. Hitler's cold and calculating interpretation of social Darwinism—according to which *survival of the fittest* meant Germany's gene pool needed to be scrubbed clean of Jews, homosexuals, the mentally retarded, and other allegedly inferior individuals—largely motivated his violence, which was responsible for what most historians agree was the second single bloodiest holocaust in history, after Stalinism. And don't forget, it took no small amount of IQ for Hitler's henchmen to dream up their insanely efficient Final Solution.

Are any of these malicious uses of IQ good reasons for eschewing logic and human ingenuity? Of course not.

The gloomy reality is, we routinely commit atrocities in the name of just about everything, including scientific knowledge, medical treatments, political utopias, you name it. For example: Dr. Josef Mengele's gruesome experiments on twin children, the U.S. military's syphilis studies on African-American men, scientists' radiation studies on mentally challenged children, CIA-sponsored tests of LSD on unsuspecting prisoners, and Joseph Stalin's wholesale slaughter of the millions who opposed Soviet communism.

Evil behavior appears to be a cursed and universal feature of the *Homo sapiens* experience. Science doesn't have any satisfactory explanation for why this is so, its endless nature-nurture debate notwithstanding. Fortunately, God does.

God not only describes our evil impulses to a T, He offers us a practical way for bringing them under control: a loving relationship with Him. It's a way, I venture to say, that's done more to help people overcome colossal problems—turn their lives around by 180 degrees—than psychologists, sociologists, and all other scientific experts combined.

Take the life of Chuck Colson, for example, President Nixon's infamous "hatchet man." In 1973, at the height of the Watergate scandal, Colson converted to Christianity. Two years later, following his imprisonment in an Alabama jail, Colson founded Prison Fellowship Ministries (PFM); and the rest, as they say, is history.

In a 2003 University of Pennsylvania report on faith-based prison programs, sociologist Byron R. Johnson and the late David B. Larson, a psychiatrist, describe the outcome of Colson's profound religious transformation as "easily the largest organized prison ministry in the United States." According to PFM's 2002–2003 annual report, *Fruit of the Vine*, the ministries now reach millions of inmates, ex-inmates, and their families annually throughout the U.S. and 104 foreign countries. All thanks to the hundreds of thousands of volunteers who donate their time because they, like Colson, believe genuinely in what Jesus says in Matthew 25:40: "I tell you the truth, whatever you did for one of the least of these brothers of mine, you did for me."

The story doesn't end there. One of PFM's volunteers, Mary Kay Beard, is a convicted bank robber who was once on the FBI's Most Wanted list. Talk about having one's life turned around by God! After giving her life completely to Him and joining PFM, Beard founded Angel Tree, which ministers specifically to the children of inmates and ex-inmates. Angel Tree's stated goal is to help break the cycle of crime within families.

Says Beard, with a humility so very characteristic of high-IQ/high-SQ persons: "Angel Tree was not my project; it wasn't even my idea. It was God's idea. He just allowed me to be the instrument that He used to plant the seed."

Imagine

Still, let us imagine for a moment: what if our planet somehow managed to rid itself of God? Are atheists correct—would we all be better off?

In one of his most famous songs, the late ex-Beatle John Lennon asks us to imagine just that—a world in which there are no countries, nothing to kill or die for, and no religion, too. In this imaginary world, Lennon asserts, surely everyone would live in peace and harmony.

For the record, I love the song; every time I hear it, I res-onate with Lennon's yearning for a better future. But really: No countries? No religion? Nothing to kill or die for? You might as well imagine people having no brains, no choices, no freedom of expression. Lobotomy, anyone?

Truth is, if there weren't anything to kill or die for, there wouldn't be anything to live for, either. Like it or not, so long as people are in charge of their own lives, there will be diver-sity; and so long as there is diversity—and let's pray there always is—inevitably there will be differences of opinion.

Remember Richard Harris as Bull McCabe in that 1990 movie *The Field*? For many a long and hard year, so the story goes, McCabe tenant-farms a tract of land owned by a widow who one day decides to put it up for auction, just to spite him. Everyone in the small Irish town recognizes the injustice and plans to back off and let "the Bull" win the auction uncontested; but a young, arrogant American stranger arrives on the scene and decides to bid against him.

That's when Richard Harris delivers what I consider the most memorable lines of the entire movie. He stands eyeball to eyeball with the cocky stranger and growls these words: "My advice is to stay out of it. This is deep. Very deep. Deeper than you think."

When the stranger fails to heed his advice, the Bull takes him on. Bull McCabe is willing to put his life on the line, not just for the land he's worked so long and hard to improve, but to challenge what he perceives is a grave wrong.

Moral of the story? In life, there are certain things we take much more seriously than others, such as justice, liberty, love, privacy, security, property, and, yes, religion. Indeed, most of us not only *live* for these things—for the right of an accused individual to have a fair trial, for the right of the Rosa Parkses of the world to sit anywhere on the bus they please, for the love of our family and friends, for the freedom to worship however, whenever, and wherever we wish—most of us are willing to die for them as well. Even atheists.

I discovered this irony when I was in Iceland, filming for *Good Morning America*. My soundman was a devout atheist in his early twenties. During our chats, he offered up the usual complaints about religion, including the biggie: "The world would be better off without it," he said, "considering all the violence that's been done in God's name." That's when I asked him, "So, would you fight someone who tried to impose a belief in God on you?

Would *you* be willing to fight and die for your atheism?" He didn't even blink an eye when he told me, yes, he most definitely would.

The irony is even more dramatic when we look at the horrific violence Marxism, Stalinism, and other godless political regimes of the twentieth century have done. In a nine-hundred-page book titled *The Black Book of Communism: Crimes, Terror, Repression*, authors Stephane Courtois, et al., document in bloodcurdling detail just how awful Communist governments really have been.

According to conservative estimates, from 1917 to 1991, atheist governments massacred close to 100 million people! The three worst offenders were:

- USSR: 20 million killed

- China: 65 million killed

- North Korea and Cambodia: 2 million killed in each

The savagery of these governments, the authors explain, was not just limited to murder:

Communism has committed a multitude of crimes not only against individual human beings but also against world civilization and national cultures. Stalin demolished dozens of churches in Moscow . . . Pol Pot dismantled the Phom Penh cathedral stone by stone and allowed the jungle to take over the temples of Angkor Wat; and during Mao's Cultural Revolution, priceless treasures were smashed or burned.

It's true these Communist regimes didn't kill in the name of just their atheism; they killed in the name of their godless political philosophies. But let's not mince words. Atheism was a fundamental part of their creeds. Generally speaking, these Communist governments denounced religion (e.g., Karl Marx accused religion of being the opiate of the people) and targeted much of their hostility specifically at churches, temples, priests, ministers, and rabbis, together with men, women, and children who refused to renounce their faith in God.

Imagine a world without God? Imagine a world where IQ reigns supreme? We don't need to. We've seen the reality of such worlds, and it's very far removed from the romanticized fantasy memorialized in Lennon's song.

To reiterate: these atrocities—the worst ever recorded in the history of our species—were committed by Communists. They were masterminded and executed not in the name of God, but in the name of godlessness. SQ had little or nothing to do with it. It was all about atheists bent on creating a thoroughly rational, material world. It was all about IQ.

These unparalleled acts of savage, wanton violence committed by godless individuals make liars of atheists who accuse God and religion of being the main sources of bloodshed in the world. They're not even close to being correct.

Religion a backward superstition? A comforting delusion? Given the utter fiasco of godless political regimes, the only backward superstition here is the belief that atheists are somehow smarter than those who believe in God. And given the personal, social, and cultural devastation ruling atheists have inflicted on people, the only comforting delusion here is the belief that atheists are somehow more advanced, more enlightened than people who believe in God.

Imagine that.

FOUR

The Gift That Keeps on Giving

> I do not feel obliged to believe that that same God
> who has endowed us with sense, reason, and intellect
> has intended us to forego their use.
>
> —Galileo Galilei

Rainbows appear at the tail end of most rainstorms. But *why*? During my years of teaching physics at Harvard, I loved asking my students that question, together with this one: if it rained *plastic* instead of water, what effect, if any, would that have on the appearance of rainbows?

I'll give you the answer in a moment. But first, I wanted to explain that rainbows occupy a very special place in my heart. They represent why I love science so much, but also why science is not my object of worship.

Anyone can look at a rainbow and *ooh* and *aah* at its beautiful colors. But I can look at a rainbow and appreciate it at a much deeper level, all because my scientific education has clued me in on the unspeakably eloquent laws of light that explain how a rainbow's colors are produced.

As a scientist, for example, I know that:

- Rainbows appear at the end of rainstorms because it is then that you have the two prerequisites for making them: 1) water droplets suspended in the sky and 2) sunlight.

- A rainbow is created when sunlight enters millions upon millions of water droplets, bounces off their back surfaces, and upon exiting fans out in different colors and different directions—like kids at recess fanning out onto the playground. (Technically speaking, that sequence of events is called *refraction*, *reflection*, and *dispersion*.)

- A rainbow's visible colors are always arrayed in the same order: red, orange, yellow, green, blue, indigo, and violet. Beyond the red and violet are invisible colors called *infrared* and *ultraviolet*, respectively.

- The best way to see a rainbow is to turn your back to the sun and raise your eyes to an angle of forty-two degrees with respect to the sun's angle above the horizon.

- Rainbows are actually circular. They appear to be arches (i.e., half-circles) only because their bottom halves are cut off by the ground you stand upon. If you wish to see them in their full circular glory, as I have on many occasions, you need to view them from high above the ground, such as onboard an airplane.

You probably had no idea there was so much to know about rainbows! I didn't either—until I became a scientist.

Which brings me to that question about plastic rain. Here's the answer: Sunlight interacts differently with plastic than it does with water. (Technically speaking, plastic and water have different *indices of refraction*.) When sunlight enters a clear polystyrene droplet, the fan of colored lights that emerges does so at an angle of nineteen degrees, less than half of the normal forty-two degrees.

Consequently, if it ever rained plastic—for example, on some strange planet—rainbows would arch into the sky less than half as high as they do here on earth. Such rainbows would be lower and smaller, but in every other way just as colorful and beautiful!

How Science Came to Be

Before there was science to regale us with fascinating insights into how the natural world operates, there was folk wisdom. People took note of certain interesting peculiarities about their everyday world and turned them into rules of thumb.

Tree crickets chirp faster on warm nights than cool ones, so ancient peoples surmised a rule of thumb that today can be stated this way: to know the air temperature in degrees Fahrenheit, simply count up the number of chirps you hear in fourteen seconds and add forty.

No one had a clue why such rules were true, and no one particularly cared. All that mattered was they worked. Like magic.

Today, the *Old Farmer's Almanac* is full of such prescientific folk wisdom:

A) If the sun in red should set,
   The next day will surely be wet.

B) When the stars begin to huddle,
   The earth will soon become a puddle.

C) Clear moon,
   Frost soon.

All three maxims are true, and farmers today still live by them. But it wasn't until science came along that we learned the reasons *why*, namely:

A) The high, gossamerlike clouds that usually precede rainstorms act like a sponge, soaking up all the colors contained in sunlight, except for red.

B) At night, those same invisible clouds tend to smear out the light from a single bright star, making it look like a small, hazy cluster.

C) Frost happens most often during cold nights when there's no wind or warmth to stir up the air, thus causing the moon to look crystal clear.

Before science, there was also armchair philosophy, the opposite of folk wisdom. Armchair philosophers looked inward, not outward, relying solely on their minds—their common sense—to figure out how the world works.

When people wondered why rocks always fell downward and never upward, common sense led armchair philosophers to theorize it was because rocks naturally desired to reunite with their original source, the earth. *Big* rocks fell *faster* than small ones because they sought out Mother Earth with greater affection.

For thousands of years, nobody seriously thought to put such theoretical notions to the test. Why bother? They had to be true. They were too logical not to be.

Enter Galileo Galilei and Isaac Newton. Four centuries ago, in a stroke of genius, they married folk wisdom with armchair philosophy—united the power of observation with the cleverness of theory—and *voila!* The scientific method was born.

The scientific method is our best recipe yet for getting at physical truths. There's no one official version of it, but here's what it basically boils down to:

1. *Observe*. Look around and take careful notes of what you see, trying at all times not to let emotions, prejudices, or weaknesses corrupt your observations.

2. *Hypothesize*. Try hard to make sense out of your observations. Do they seem to follow some consistent rule? If so, then try putting it into words (or equations). It's what's called your *hypothesis*—a fancy word for educated guess.

3. *Test*. Put your hypothesis on trial by coming up with clever ways of testing it, either in a laboratory or out in the real world.

4. *Discover*. Depending on the verdict of your experiments, you'll discover (a) your hypothesis was completely on target, or far more likely; (b) your hypothesis is faulty—its predictions aren't reliable enough.

5. *Correct*. If your hypothesis came up short, you need to fix it. After that, you need to go back to step 3 and test it all over again—and so forth, until you get a hypothesis that appears to be 100 percent reliable; at that point it qualifies as a *scientific law*.

Needless to say, science can be a tedious business, requiring loads of patience and perseverance. But much of the time it's worth the effort. Since its invention centuries ago, the scientific method has enriched our lives in countless ways.

• In politics: James Madison's famous system of checks and balances for the U.S. Constitution was inspired by Newton's third law of motion—for every action, there's an equal and opposite reaction.

• In art: the psychedelic-like Impressionist movement was inspired by Newton's discovery that white light, far from being pure, actually consists of every color of the rainbow.

• In space exploration: Newton's law of gravity enabled

us to break free of the earth's grip and navigate safely to the moon's surface.

- In popular literature: Sir Arthur Conan Doyle's Sherlock Holmes used the scientific method to solve crimes that left other detectives scratching their heads.

- In advertising: the slogan "Scientifically Proven" has become the ultimate seal of approval.

- In medicine: scientific advances in sanitation, nutrition, pharmaceuticals, vaccinations, and surgery have helped us to stave off death and disease.

- In religion: the scientific method has helped to clarify our understanding of God and His creation. For instance, most of us now realize thunder is caused not by blows of some divine hammer, but the implosion of air. That congenital diseases are caused not by the sins of our fathers, but by their faulty genes. That pregnancies are caused not by the visitations of a lunar god on women, but the union of egg and sperm.

How Science Came to Me

I could go on and on. There's a lot to admire about science; that's why I love it. Indeed, the story of how I came to be a theoretical physicist is like a fairy tale—a story about love at first sight.

Once upon a time, when I was about nine, Dad came home with a paperback book titled *Nuclear Forces*. He'd found it at a sidewalk sale on one of those mark-down tables and figured I might find it interesting—not realizing it was a graduate-level college textbook!

Not knowing any better myself, I cracked it open, and miracle of miracles, over the course of months and years, I actually began cobbling together some of its essential concepts: nuclear spin, parity, Bose-Einstein statistics, fermions, et cetera, et cetera, et cetera. Just like that, God had begun to plant seeds in my life that one day would grow into a full-fledged, formal education in theoretical physics.

But there was a hitch. Compared to the lofty physics I was studying on my own, elementary school was unbearably boring; so I became the class clown, the kid who just couldn't sit still or keep his silly mouth shut.

At first, my disruptive behavior went unpunished, maybe because the teachers knew I was a serious student at heart. But finally in the fifth grade, my teacher put his foot down. He marked my report card with straight S's (for satisfactory), except for one very glaring and onerous N (for nonsatisfactory). To this day, I remember all too well what that awful N was for: *inability to exert self-control*. Those were the exact words.

The principal summoned Mom and Dad to his office, which, besides being humiliating for them and me, led to a scolding that was only partially successful. I just couldn't help myself.

In middle school, my math teacher took to calling me Michael Jillion, because of my insatiable and disruptive curiosity. I was constantly interrupting him with a "jillion" annoying questions.

In high school, my physics teacher, Mr. Wiley, recognizing

how bored I was, excused me from having to sit in class. Instead, he let me putter around in the back room, where I did my own experiments and helped him write tests for the other kids.

During all this strangeness, my saintly parents continued to encourage my scientific proclivities. They bought me a telescope, allowed me to build a full-fledged chemistry lab in the garage, and stood by without interfering when I blew up rockets, dug up worms, and revved up noisy engines—all in the name of science.

They even held their peace when I discovered a simple way to make hydrogen gas—which is buoyant but extremely explosive—and used it to inflate balloons to which I tied little capsules manned with bugs; I wanted to see how well the critters survived at high altitudes.

As if all that weren't enough for my poor parents to handle, one day, while I was still in school, God sprang on me yet another extravagant ambition, one even more unexpected and puzzling than my wish to become a scientist. In reading the newspaper, I'd noticed there were regular columns on nearly every subject under the sun—sewing, bridge, stamp collecting—but none whatsoever on science.

After stewing over this indefensible omission, I worked up the nerve to call Richard Feynman, the famous physicist at nearby Cal Tech. But he was out, so I left a message, even though I figured he'd never return a call from a complete nobody like me.

A few days later, I was at home when the phone rang. Mom answered, and I remember her being confused about whoever was calling and almost hanging up. She put her hand over the mouthpiece and called out to me that some man named Fainman or Feinmon wanted to talk to me. I was thunderstruck!

After rushing to the phone and pausing a moment to collect myself, I finally managed to explain to Feynman—trembling in my shoes the whole time—my belief that there should be a weekly science column, and he was the perfect person to write it. Feynman was a man of few words, and I'll never forget how he replied: "I don't have the time, kid, but if you feel that strongly about it, why don't *you* write the column?" Then he hung up.

His suggestion sounded far-fetched to me, but still, I typed up some sample articles, tested them on my younger sister Debbie, then mailed them out to editors nationwide. All of them rejected me. Nevertheless, I didn't give up.

Finally, some months later, I managed to convince the editor of a chain of weekly community newspapers to publish my columns for free. Within a few years, they became such a hit, the editor began paying me a whopping five dollars per article.

On the academic front, the ups and downs continued. After surviving high school, I matriculated to UCLA, where I immediately landed a coveted UCLA President's Undergraduate Fellowship and my very own lab on the top floor of the physics building—which, I might add, had a million-dollar view of Los Angeles.

After graduating from UCLA with honors, I went to Cornell to study physics, becoming the first in my family ever to leave the Southwest or attend an Ivy League school. But once again, there were hitches.

Within weeks of my arrival at Cornell: my girlfriend from UCLA broke up with me, my professors wouldn't stop ridiculing Los Angeles, and my mother was diagnosed with breast cancer. (Years later, I discovered Mom first felt a lump in her breast around the time I was accepted into Cornell. She deliberately put off going to the doctor, fearing the diagnosis might cause me to change my plans and thereby ruin my dream of becoming a scientist.)

At Cornell, my lifelong pattern of nonconformity persisted. Instead of sticking to my studies like a well-behaved grad student, I took summers off to intern at *Science News* magazine in Washington, D.C.; founded The Leonardo da Vinci Society, whose purpose was to encourage a dialog between the sciences and humanities; and asked for the university's permission to broaden my major to include astronomy and mathematics.

On account of all these eccentricities, I became the subject of numerous physics faculty meetings at which the question was always the same: "What in the world do we do with Michael?" My behavior just didn't fit their notion of normal. But thanks to God's providence—in the persons of some broad-minded professors—I managed to avoid being kicked out of school and to graduate with a PhD in not just one discipline but three: physics, math, and astronomy.

Like any good fairy tale, my love affair with science had a happy ending—an *incredibly* happy ending. Not only did science make it possible for me to end up at Harvard and ABC-TV, it enabled me to travel all over the world and meet some of its most interesting, intelligent, and influential people. Not bad for a kid from East Los Angeles!

If I were an Intellectual Cyclops, this is where I'd gush that I owe *everything* to science. But I'm not, and I can't. In this respect, too, I'm different.

For many of my academic colleagues, science is more than a profession or calling; it's a religion called Scientism. I'm referring here to a particular type of godlessness about which even other atheists are wary—for instance, Uncertain Atheist Michael Shermer. In his book *How We Believe*, he remarks skeptically: "Scientism is their religion, technocracy their politics, progress their God. They hold an unmitigated confidence that because science has solved problems in the past, it will solve all problems in the future."

For my Scientism colleagues, the scientific method is more than a powerful recipe for understanding the natural world; it's the Scientific Method (with capital letters), *the* most powerful recipe for understanding *anything* and *everything*. The Scientific Method is as sacred to them as the Bible is to born-again Christians.

Not for me. I love science—as you've seen, I've devoted my entire life to it—and science loves me. Science has provided me with decades of intellectual satisfaction and spiritual joy. But that's where it stops.

I revere the scientific method, but I don't worship it. I have faith in science, but as the famous Harvard psychologist William James once observed, "Faith means belief in something concerning which doubt is theoretically possible." At the end of the day, despite all of science's spectacular successes, I doubt seriously its ability to lead us unescorted to ultimate truth.

I'm reminded of this feeling every time I look at a rainbow. As a scientist, I can appreciate the colorful bows in ways that laypeople can't possibly imagine, ways they're simply unable to see with their scientifically undeveloped perceptions of reality.

But also, as someone with a stereoscopic faith in God, I can appreciate rainbows in ways even my Scientism colleagues can't possibly imagine, ways they're simply unable to see with their *spiritually* undeveloped perceptions of reality.

By being familiar with Genesis 9:16, I'm able to appreciate rainbows as stunning symbols of God's immortal affection for us: "Whenever the rainbow appears in the clouds, I will see it and remember the everlasting covenant between God and all living creatures of every kind on the earth."

That's why, whenever I see a rainbow, I not only take pleasure in its colors and wonder at the elegant optical mechanisms that create them, I absolutely marvel at the extraordinary genius—talk about IQ!—of its Author and Finisher.

I love science, but most of all, I love God—for creating a universe that includes rainbows *and* a scientific method that seeks to appreciate them and everything else in the cosmos on a deep, subsurface level. As Pope Pious XII stated in his 1953 Christmas message: "The church welcomes technological progress and receives it with love, for it is an indubitable fact that technological progress comes from God, and therefore, can and must lead to Him."

Yes, I love science, but it's not my god. Instead, for me, science will always represent a precious gift *from* God.

FIVE

Turning a Blind Eye

Every man takes the limits of his own field of vision
for the limits of the world.

—Arthur Schopenhauer

A 3-D movie or comic book usually comes with a special pair of glasses. Years ago, the lenses were made of two opposite-colored cellophane filters, one red, one green. Nowadays it's more common for the lenses to be made of oppositely polarized filters, analogous to venetian blinds: one with its invisible "slats" oriented horizontally, one with them oriented vertically. In any case, the lenses of 3-D glasses are always *opposites* in some way.

If you look at a 3-D movie or comic book without those special glasses, or through only one of its lenses, then forget about it: you'll see nothing but a tangle of flat, unrealistic-looking images. But put them on, and *amazing!* You'll see images that are so true to life you'll feel you can reach out and touch them.

I claim IQ and SQ are like the lenses of those 3-D glasses. For one thing, they're opposites. And for another, if you use

67

them both, *wow!* You'll suddenly see so much depth, beauty, and meaning in your life and the universe, I promise you'll actually feel the presence of God.

With stereoscopic faith, you'll see reality in its full, multi-dimensional glory: space and time on the one hand, meaning and purpose on the other. I mean that literally. I believe *meaning* and *purpose* are actual dimensions, no less so than *space* and *time*.

If, however, you're an Intellectual or a Spiritual Cyclops—someone who isn't practicing stereoscopic faith—then I believe you'll always be blind to reality's true profundity. As Isaiah explained, you'll always be "ever hearing, but never understanding . . . ever seeing, but never perceiving" (6:9).

You'll be like the native who thinks communicating with drums is high technology yet is clueless about the radio waves whizzing over his head, carrying important information among the peoples of a planet he's never explored. How do you even begin explaining any of that to him?

Worse still, if you're an Intellectual or a Spiritual Cyclops, then your blind spots can cause you to have certain ugly prejudices. Just listen to Intellectual Cyclops Madalyn Murray O'Hair, the American atheist who successfully opposed prayers in schools. Referring to people who believe in God, she said: "I feel everyone has the right to be insane."

The prejudices of a Cyclops are exactly what Matthew 7:3, 5 warn us away from having:

Why do you look at the speck of sawdust in your brother's eye and pay no attention to the plank in your own eye? . . . You hypocrite, first take the plank out of your own eye, and then you will see clearly to remove the speck from your brother's eye.

Apart from being reprehensible, such prejudices can be incredibly passionate: Quite literally, a Cyclops just can't see why others don't agree with him. As the eighteenth-century polymath Georg Christoph Lichtenberg, one of history's great thinkers, put it: "With most people, unbelief in one thing is founded upon blind belief in another."

What follows are some of the Cyclopes' most common one-eyed prejudices. When you read them, no doubt you will be quick to recognize them in others. But also try to be honest, even if it's painful. Do you see yourself in any of them? Are you a person who relies mostly on *blind faith*? Or are you someone who relies mostly on *blind reason*? If so, then perhaps you're a Cyclops who needs to think about getting fitted for some prescription eyeglasses—of the 3-D variety.

If you are neither, then congratulations! You're one of those rare individuals who already sees life and the universe with stereoscopic faith, a perspective that's perfectly balanced between IQ and SQ.

Spiritual Cyclopes

Seventeenth-century church officials—Spiritual Cyclopes—refused to look through Galileo's crude little telescope because they didn't see any point to it. Why waste precious time in dulging some misguided heretic who believes the earth revolves around the sun? Yet, as we all know, the physical evidence vin-dicated Galileo and humiliated the church officials.

Today, despite the humbling lessons of history, there are still many religious people who deliberately turn a blind eye to the physical world. Indeed, modern-day Spiritual Cyclopes point to the Bible's Doubting Thomas story to argue that blind faith is a cardinal virtue, that a pure, unquestioning, unthinking belief in God is somehow superior to any other kind.

In particular, they cite this passage in John 20:29: "Jesus said to him, 'Thomas, because you have seen Me, you have believed. Blessed are those who have not seen and yet have believed'" (NKJV).

Spiritual Cyclopes also defend their one-eyed approach to God by referring to an incident described in Matthew 18:2–3: "[Jesus] called a little child and had him stand among them. And he said: 'I tell you the truth, unless you change and become like little children, you will never enter the kingdom of heaven.'"

To Spiritual Cyclopes, this sounds as if Jesus is commanding us to become childishly, blindly trusting—especially since the verse right after it says, "Therefore, whoever humbles himself like this child is the greatest in the kingdom of heaven."

To Spiritual Cyclopes, intellectualism breeds pride, the very antithesis of humility. Therefore, they reason, in calling us to become humble children, Jesus was effectively saying that SQ, not IQ, is our ticket to heaven.

The icing on the cake for Spiritual Cyclopes is found in pas-sages such as Matthew 11:25, where once again Jesus appears to be disparaging IQ: "Jesus said, 'I praise you, Father, Lord of heaven and earth, because you have hidden these things from the wise and learned, and revealed them to little children.'"

For the record, I agree with some of what these Spiritual Cyclopes are reading into the Bible. We all know of individuals who don't wear an academic education very well. Their intelligence goes to their heads—no pun intended—like the college freshman who's thunderstruck at how much dumber his parents have gotten since he was a kid.

But it need not be that way. Folks, I've seen it and probably you have, too: it's possible for a person to be intelligent and humble at the same time.

In my opinion—and here's where I part company with Spiritual Cyclopes—the Bible asks us to become like children and avoid the sin of pride not by remaining ignorant about the physical world, but by striving to develop a faith that is stereoscopic, a faith that involves our entire beings. Nowhere is this commandment more clear, I believe, than in Mark 12:30: "And you shall love the LORD your God with all your heart, with all your soul, with all your mind, and with all your strength" (NKJV).

In 1 Thessalonians 5:21, Saint Paul exhorts us: "Test everything. Hold on to the good." Again, this is clearly *not* an invitation for us to become Spiritual Cyclopes, to turn a blind eye to the natural world. Just the opposite.

I believe the Bible here is urging us to observe every little thing that goes on around us—every claim, every phenomenon, every impulse—and to judge each one critically, sorting out the trustworthy ones from the bogus ones. In other words, the Bible is commanding us to behave like *scientists*!

How very appropriate: God calling us to become both children *and* scientists. Kids are natural-born scientists, always asking questions about how the world works. "Why is the sky blue?" "Where do birds go during a storm?" "What is electricity?" According to God's own words, that's how we're to be as adults, too: cognizant of all the wondrous things He's created for our intellectual and spiritual edification.

Please don't mistake this for a call to become worldly. In James 4:4, God makes it clear that being infatuated with the imperfect world we've created for ourselves is definitely not the way to go: "Don't you know that friendship with the world is hatred toward God? Anyone who chooses to be a friend of the world becomes an enemy of God."

Rather, God is urging us to avoid becoming Spiritual Cyclopes; to appreciate the natural world, *His* creation, as a powerful way of coming to know Him more fully. That's the bottom line. As Saint Paul tells us in Romans 1:20: "For since the creation of the world God's invisible qualities—his eternal power and divine nature—have been clearly seen, being understood from what has been made."

Intellectual Cyclopes

Intellectual Cyclopes can be just as close-minded as their Spiritual counterparts—often in the name of being skeptical, which is a shameful fraud. Genuine skepticism, I believe, requires us to approach heretical ideas with open minds.

Here's an example of what I mean. Back in the 1960s,

biologist Lynn Margulis's personal study of microbes led her to a thoroughly novel hypothesis: contrary to orthodox Darwinian reasoning, she proposed, early life evolved through *cooperation* rather than competition.

Her fellow scientists had a right, even an obligation, to be skeptical—Margulis's ideas were pretty far out. But for Intellectual Cyclopes, skepticism always translates into close-mindedness—and worse.

When Margulis solicited the National Science Foundation (NSF) for research funding, for example, she was reportedly subjected to intense scorn. Recalls Margulis, "I was flatly turned down." And the grants officers added 'that I should never apply again'" (*Boston Globe*, June 22, 1987).

To Intellectual Cyclopes at the NSF and elsewhere, Margulis's suggestion that the theory of evolution might be incomplete or incorrect was as inconceivable as color is to someone who's congenitally sightless. That's why the NSF refused Margulis's request for funding. Why waste precious resources putting her idiotic hypothesis to the test? *Why even bother looking through the telescope?*

Notwithstanding this irrational opposition, Margulis persevered, and today her so-called *symbiotic theory* is widely accepted by her peers. In 1983, she was even elected to the National Academy of Sciences, an honor considered second only to receiving the Nobel Prize.

The intellectual dogmatism (what I've heard called *pathological skepticism*) that arrogantly dismissed Margulis is the same blinkered mentality that denies the existence of God out of hand. As far as Intellectual Cyclopes are concerned, there's no reason whatsoever for any serious-minded person to entertain the possibility of a spiritual realm: the physical world is clearly all there is. It's *everything*.

No wonder many of today's Intellectual Cyclopes speak rather pretentiously about the scientific quest for *the theory of everything*. For example: Stephen Hawking's recent book by that name and astrophysicist John Gribbin's *The Search for Superstrings, Symmetry, and the Theory of Everything*.

What these men are speaking of here is actually the search for a single coherent theory that can replace today's hodge-podge of hypotheses. Scientists hope the fabled theory of everything will be able to explain, under one roof, all the physical forces known to science: gravitational, electromagnetic, nuclear, and weak. They're like an interior decorator aching to refurnish a house with pieces all made from a single style.

It's a worthy goal, but the ideas such scientists offer are far from being a theory of everything—or even a theory of much of anything, as far as what matters most to people. For example, such a theory wouldn't explain even the simplest things about human existence: why two strangers fall in love; why catching sight of a beautiful sunset can cause us to stop dead in our tracks and gawk; why something deep within us resonates, like a tuning fork, with the feeling that we're on this earth for a reason.

Putting it metaphorically: the music of life will never be found in any theoretical analysis of its notes or sounds. Rather, it will be found nowhere and everywhere throughout our conscious being—something the Intellectual Cyclopes' theory of everything can't even begin to comprehend, considering their wish to reduce everything down to the random machinations of subatomic particles and forces.

In his classic book *My View of the World*, Nobel Prize–winning physicist Erwin Schrödinger, the legendary coarchitect of quantum theory, explained it this way:

> [Science] is ghastly silent about all and sundry that is really near to our heart, that really matters to us. It cannot tell us a word about red and blue, bitter and sweet, physical pain and physical delight, knows nothing of beautiful and ugly, good or bad, God and eternity. Science sometimes pretends to answer questions in these domains, but the answers are very often so silly that we are not inclined to take them seriously.

Most disturbing of all is the ghastly silence from the so-called human behavioral sciences—psychology, sociology, anthropology—concerning God. I say *so-called* because it's inconceivable to me anything can be thought of as a "human science" that is so inept and aloof in dealing with our species' nearly universal belief in a supreme being. Allow me to elaborate.

The Bedeviling G-Factor

There's a popular 1970s TV game show you can still catch in reruns called *To Tell the Truth*. During each half-hour program, panelists grill three contestants who all claim to be Mr. X, someone in an interesting line of work. At the end of each show, the panelists vote, the real Mr. X stands up, and everyone applauds.

Today, many people are under the impression this is how the game of science is played. Scientists interrogate Mother Nature with controlled experiments, dream up various hypotheses to explain the data, then wait until some decisive piece of evidence shows up to settle things, and everyone applauds.

Too bad that's not actually how it works. In reality, there's always more than one way to explain the existing data (see chapter 7), so there's never a climactic moment when the real Mr. X stands up. Instead, science is always having to deal with pretenders—some more plausible than others—who are constantly half-rising from their seats, then sitting back down again. It's like some big, interminable tease.

How does science cope with this steady parade of would-be theories? One way is through the use of a tough-minded weeding-out principle credited to the fourteenth-century philosopher William of Occam. It's called Occam's Razor, and put simply it says this: whenever you're faced with a variety of candidate hypotheses, each of which explains the data equally well, always favor the one that explains things most efficiently, based on the fewest number of assumptions—especially assumptions that sound far-out or aren't amenable to scientific testing.

Mind you, it's a completely arbitrary rule. Early scientists could've just as easily adopted a rule that favored, say, the least mathematical theory, or the most commonsensical theory, and so forth. But Occam's Razor seems to work, so for now scientists are sticking to it.

Please note, however, that in its blind allegiance to simple-mindedness, Occam's Razor automatically eliminates any hypotheses involving the G-word, a fact immortalized in the story of the famous French mathematician-astronomer, Pierre Simon de Laplace.

In 1829, Laplace proudly handed Napoleon his *magnum opus*, a five-volume treatise on how the universe works titled *Celestial Mechanics*. After looking it over, Napoleon is said to have remarked with surprise: "You have written a large book about the universe without once mentioning the author of the universe." To which Laplace reportedly replied, "Sire, I have no need of that hypothesis." (*"Je n'ai pas besoin de cet hypothese."*)

As an aside, I can't resist mentioning that given the increasingly complex and mystical (one might even say supernatural) hypotheses astronomers are taking seriously these days——multiple universes, imaginary time, dark energy, and so forth——the much-maligned hypothesis of God is emerging as the least bizarre of all. Scientists will never be able to admit it, for reasons I'm about to explain, but it's amusingly ironic that given the peculiar new direction astronomy is taking, God is fast becoming the much-sought-after *simplest explanation of all!*

For now, the point is: whilst billions of people on Planet Earth warmly embrace God, Mr. Occam doesn't. For him, God is too SQ, too complex, too far-out, and perforce must be dismissed without any consideration. That's why God will never be a part of science's so-called theory of everything.

It's nothing personal against God, mind you. Science isn't saying God doesn't exist, nor is it saying He does. Science, by mutual agreement of its practitioners, is completely neutral on the subject. (The technical term for this automatic exclusion without prejudice is *methodological naturalism*.) That's why, for instance, whenever Stephen Hawking is asked his scientific opinion on godly matters, his answer is reportedly always the same: "I don't answer God questions." Very wise.

Science is far more comfortable juggling relatively simple concepts, such as randomness, coincidence, purposeless-ness—concepts that can be described easily enough with logic and mathematics. For that reason, science is more like the pop-ular board game Taboo. Contestants must get their teammates to guess the name of a person, place, or thing by shouting out clues, but with a catch: certain clues are taboo—players can't use them.

In science, if the question is, "How did the universe begin?" players are allowed to offer up any explanation their fertile IQs can imagine: explosion, implosion, quantum perturbation, what have you. But they can't even think of crediting *God*. In the game called science, officiated by Mr. Occam, invoking the G-word is definitely considered taboo.

The irony is, for these very same reasons, you need not ever worry that daily, headline-grabbing scientific revelations are somehow gradually evicting God from the universe. Evicting? Science can't evict God from the universe, because science has never allowed Him into it in the first place.

God will always be too large to fit into science's severely

circumscribed description of the cosmos; too spiritual for science, in its allegiance to Occam's Razor and blind reason, ever to acknowledge His existence. It's not a slam against my chosen profession to say that, mind you, merely a crucial observation about the strict, one-eyed rules it plays by.

And now I end this chapter as I began, with this question: did you recognize yourself in any of these discussions? Are you perchance either an Intellectual or a Spiritual Cyclops? If so, then please rest assured my purpose here is not to judge anyone—that right belongs exclusively to our Creator—but to be of some help.

I pray that by knowing one another's blind spots, Spiritual and Intellectual Cyclopes will be more understanding of one another's very different perspectives on the world. As the twentieth-century American journalist Alexander Chase said, "To understand is to forgive."

Finally, I pray that by recognizing how costly being a Cyclops is to any search for truth, each of us will strive to keep both eyes wide open, lest we miss seeing the full multidimensionality of the physical world—and beyond.

SIX

Faith by Any Other Name

> I think that science without religion is lame
> and, conversely, religion without science is blind.
>
> —Albert Einstein

Nowhere is the opposition between Intellectual and Spiritual Cyclopes more apparent than in their attitudes toward the word *faith*.

I'm always somewhat amused at the lengths most atheists will go to deny that faith has anything whatsoever to do with their high-IQ view of the world. Somehow in the calculus of their minds, faith is a four-letter word, something a person can't possibly have and continue feeling smart or superior:

> A man full of faith is simply one who has lost (or never had) the capacity for clear and realistic thought. He is not a mere ass: he is actually ill. (H. L. Mencken)

We may define faith as a firm belief in something for which there is no evidence. (Bertrand Russell)

Godly people will go to equally great lengths to insist that nothing *other* than faith has anything whatsoever to do with their high-SQ view of the world. For them, faith is as hallowed as a Gothic cathedral, and logic is as unwelcome to enter as the devil himself:

Faith must trample under foot all reason, sense, and understanding. (Martin Luther)

Faith would have no merit if reason provided proof. (Pope Gregory I)

In effect, the warring sides claim: a person of faith has no need of reason, and a person of reason has no need of faith. Do you believe that?

Do you believe reason and faith are opposed to one another? That science, for instance, has nothing whatsoever to do with believing in things that can't be proven? That believing in God has nothing whatsoever to do with the discoveries and achievements of science?

I don't—not for a minute. Instead, I believe that, like siblings, reason and faith look and act differently, but under the skin they share certain genetic similarities. In particular, they're identical in this regard: science and religion are each a potent mixture of both IQ and SQ.

Science Needs Faith

As a high-school student, I worked part-time for a well-known supermarket chain store. Business was slumping, so one day, a corporate bigwig who'd flown into town rounded up all of us employees for a stern pep talk. After imploring us to work harder and more efficiently, he ended with this admonition: "And remember: don't ever make assumptions about anything. You'll always end up regretting it."

Years later, when I was training to be a physicist, I learned— much to my consternation—that even *science* is not above lapsing into this risky practice. Indeed, the entire scientific high rise is founded on assumptions!

I call them Articles of Faith, because that's what they are: assumptions that cannot be proved or disproved, and that consequently every card-carrying scientist must accept solely on faith.

Here are three examples of what I'm talking about:

- The principle of sufficient reason—the assumption that every mystery has an explanation.

- The principle of the excluded middle—the assumption that every imaginable proposition is either true or false; there's no in-between.

- The principle of Occam's Razor—the assumption that the *simplest* explanation of anything is always the best.

These three Articles of Faith have served scientists well for a very long time; that's why my colleagues and I continue to believe in them quite religiously. But now and then, scientists have put their faith into an assumption that, as the supermarket guy warned, ended up making them regret it. Like this famous example:

- The principle of parallel lines—the assumption that two infinite, parallel lines will never meet.

For thousands of years, scientists believed in the truth of the parallel line principle, even though no one had ever gotten around to proving it. Perhaps you yourself learned this precept in high school, never realizing it wasn't a fact, but merely an *assumption*. But then again, why would it even occur to you to suspect anything? Of course parallel lines never meet—what could be more obvious?

More than two thousand years ago, even the fabled mathematician Euclid believed in the parallel line principle; he made it one of the foundation stones of his geometry. Much later on, so did the great seventeenth-century scientist Isaac Newton; the parallel line principle is at the heart of his physics.

But then came the rude awakening.

During the early 1800s, a rapid-fire succession of eminent mathematicians around the planet—János Bolyai in Austria, Carl Friedrich Gauss in Germany, and Nikolaus Lobachevski in Russia—each independently made a discovery that utterly demolished our age-old faith in the parallel line principle.

They discovered it was possible to imagine worlds wherein

parallel lines *do* meet, *do* come together, *do* intersect! In the process, they even discovered hypothetical worlds in which parallel lines *diverge* from one another! This was mind-boggling news, akin to our first discovering the earth is round, not flat.

At first, intellectuals were able to take comfort in knowing that the worlds in which parallel lines converged or diverged were strictly hypothetical. They clung to their faith that the venerable parallel line principle was still true for our world, the *real* world.

But in the 1920s, Einstein's theory of general relativity pulled even that rug out from under them. According to Einstein's radical theory, even in *our* world—our very own universe—parallel lines behave strangely. Einstein wasn't able to say right away whether the lines converge or diverge (today's cosmologists are still debating the issue), but he did show they aren't constrained by the old principle of parallel lines.

Despite such chastening experiences to their discipline, scientists today still keep the faith. Faith that every mystery has an explanation. Faith that every imaginable proposition is either true or false. Faith that the simplest explanation of anything is always the best.

And why shouldn't they? Even though my scientific brethren place their faith on a wrong principle now and then, and live to regret it, science as a whole works—the technological evidence of that is all around us. As a dear engineer friend of mine put it, "I believe in science, because I see planes flying in the air every day!"

Religion Needs Science

During the sixteenth century, many God-fearing people insisted that Copernicus was mistaken in believing the earth spins on its axis and also revolves around the sun. They staked their faith in God on it. After all, in Psalms 93:1 it says clearly: "Surely the world is established, so that it cannot be moved" (NKJV).

We all know how that one turned out.

More recently, when I was a kid, I remember Dad telling me there were godly people who believed humans would never land on the moon. They staked their faith on it—that's how certain they were. In Isaiah 55:9, they argued, God Himself tells us: "As the heavens are higher than the earth, so are My ways higher than your ways, and My thoughts than your thoughts" (NKJV).

Surely this meant the moon—being part of the "heavens"—was off-limits to us mortals. Only God, whose ways are "higher than ours," could ever possibly visit the moon. Right?

I've always wondered what became of those poor people when on July 20, 1969, Neil Armstrong planted his foot on the gray, dusty lunar surface. Their faith in the Bible must've been shattered, or at least sorely tested. For them, God had taken another blow to the chin from big, bad Science.

And yet, He hadn't really. The Bible was never on trial, nor was God. Instead, as James Carvill might say: *It's the interpretation, stupid!*

In every generation, many of us convince ourselves that God's existence utterly hinges on one particular interpretation of some sacred verse, holy relic, or scientific discovery. (According to the 2003 World Christian Database, there are about 37,000 different denominations in Christianity alone!) In those moments, because it means everything to us, the *interpretation* itself becomes a god. A false god.

We who believe in the Bible must continue to assimilate into our various interpretations of its scriptures any and all irrefutable scientific discoveries—be it Copernicus's heliocentric theory or the moon landing—or risk taking an indefensible position that undermines the public credibility of our faith, and by association, of God Himself. As Ralph Waldo Emerson once put it, "The religion that is afraid of science dishonors God and commits suicide."

Having said that, I hasten to point out something else we should also bear in mind. Even though my religious brethren have placed their faith on a mistaken interpretation now and then, religion as a whole works—the evidence of that is everywhere.

Historically, as I've mentioned before, our belief in God has tamed the worst elements of our former barbarism, including brutishness, moral chaos, political anarchy, incest, and casual homicide. It's also played a leading role in abolishing slavery, championing women's rights, and most recently—through the historic efforts of religious leaders such as the Reverend Martin Luther King Jr.—garnering civil liberties for racial minorities.

New York's famous Bellevue Hospital—oldest in the U.S.—was started in 1658 in a poorhouse supported by a church in New Amsterdam. Says Gloria Shur Bilchik, writing in the magazine *Hospitals & Health Networks*: "Until the early 20th century, almost all hospital development was the result of private donations motivated by Judeo-Christian ideals of charity, love for one's neighbors, and dedication to a ministry of healing."

According to the *NonProfit Times*, in 2002, the seven financially largest, publicly supported philanthropies in the U.S. were: The National Council of YMCAs ($4.2 billion), American Red Cross ($4.1 billion), Catholic Charities USA ($2.6 billion), Salvation Army ($2.2 billion), Goodwill Industries International ($2.1 billion), United Jewish Communities ($2 billion), and the Boys and Girls Clubs of America ($1.1 billion)—every one of which, it might surprise you to learn, have religious foundings; their origins can be traced back to men and women of great faith.

American Red Cross, started in 1881 by Massachusetts-born Clara Barton, was inspired by the original Red Cross movement, which was founded two decades earlier in Switzerland by Henry Dunant, a Calvinist who also played a leading role in the early growth of the Young Men's Christian Association (YMCA).

Boys and Girls Clubs of America, which began as the all-boys Dashaway Club of Hartford, Connecticut, was founded in 1860 by women volunteers at a Congregationalist church mission in the city's slums. Indeed, the entire boys club movement, explains Peter C. Baldwin, in the Spring 2002 issue of the *Journal of Social History*, "originated in New England from a Protestant sense of Christian stewardship."

Likewise, Goodwill Industries International was founded more than a century ago by the Reverend Edgar J. Helms, a Methodist minister in Boston's South End.

In their 2002 article, *Objective Hope: Assessing the Effectiveness of Faith-Based Organizations* [FBOs], sociologist Byron R. Johnson and his colleagues sum up the enormity of religion's social beneficence today:

> Even excluding the contributions of literally thousands of smaller faith-based initiatives and groups, the larger FBOs alone provide more than $20 billion of privately contributed funds to support the delivery of various kinds of services to more than 70 million Americans annually.

I think it's safe to say that, based on how very well each of them works, religion and science are by far our two most powerful and precious institutions. We commonly make the mistake of portraying them as being citadels of faith and reason, respectively, but as we've seen, that's not entirely accurate.

Yes, science relies heavily on reason, and religion on faith. But both exist, both work as well as they do, solely—I repeat, solely—because each relies on a potent mixture of both IQ and SQ. Remove from its practice one or the other of these, and each institution would cease to be.

In short, contrary to what the Intellectual and Spiritual Cyclopes would have us believe, a healthy and credible science will always need faith, and a healthy and credible religion will always need reason.

SEVEN

Hope Springs Eternal

> Logic is the art of going wrong with confidence.
>
> —Joseph Wood Krutch

Many atheists, especially Arrogant Atheists, are not content to live in peace with others who don't believe as they do. They feel a need to rail against God and those of us who try living according to His commandments.

I've taken an inventory of these gripes and disparagements over the years, and from what I can see, they all fall into two basic categories. Here they are, together with several examples of each:

Complaint #1: Believing in God involves certain insurmountable (read intolerable) uncertainties.

I have never seen the slightest scientific proof of the religious ideas of heaven and hell, of future life for individuals, or of a personal God. (Thomas Edison)

Never believe anything that can't be proved. (Nobel Prize–winning chemist Irving Langmuir, when asked about religion)

We must question the logic of having an all-knowing, all-powerful God, who creates faulty humans and then blames them for His own mistakes. (Gene Roddenberry, Creator of *Star Trek*)

Complaint #2: God's worshippers are imperfect.

[Christians are] a powerful subtribe of the Hypocrites, whose principal industries are murder and cheating. (Ambrose Bierce)

God is the immemorial refuge of the incompetent, the helpless, the miserable. (H. L. Mencken)

I'm offended by the flippant, sarcastic, narrow-minded ways these atheists express themselves. But guess what? In principle, I agree with their two basic complaints!

As a man for whom having a personal relationship with God is the very top priority in life, I, too, wish there were less distance and uncertainty between God and us mortals. I wish I could shake hands with God, right here and now.

Also like the atheists, I, too, wish that all of us were less imperfect than we are—that goes without saying. Without doubt, our most important asset is also our most devastating liability, namely: our God-given freedom to choose between good and evil. Or, as feminists were so fond of trumpeting in the sixties: *free to be me!*

So, yes, atheists and I are of a like mind with regard to these matters. But here, I dare say, is something that will surprise you even more: the atheists' two complaints about spirituality are also true about intellectualism: (1) believing in science, logic, and reason involves certain insurmountable uncertainties, and (2) the architects of our modern-day scientific and technological world are imperfect. Allow me to explain.

Insurmountable Uncertainties

In my formal training as a theoretical physicist and my study of the history of science, I've identified no fewer than three examples of how our IQ, like our SQ, is plagued by insurmountable limitations. Here they are.

1. Scientific Explanations Come and Go Unpredictably

During the past couple of centuries, the total number of professional scientists has mushroomed. According to the National Science Foundation's Science Resources Statistics Division, in 1999, there were nearly 11 million gainfully employed, college-educated scientists and engineers in the U.S. alone.

What's more, all of them are busy writing. In his bestseller *Megatrends*, John Naisbitt estimates that 6,000 scientific articles are written every single day. Put another way, I've heard it said that the number of scientific articles produced in just the past twenty years exceeds the total number published since the beginning of the printed word!

With all that brainpower hard at work, you might think we're becoming exponentially smarter about how the world works. But it's not so, as explained by the authors of one of my favorite books, *The Encyclopaedia of Ignorance*, which states: "Compared to the pond of knowledge, our ignorance remains atlantic. Indeed, the horizon recedes as we approach it."

It reminds me of the joke about the rich man who spends a fortune and many years searching for the meaning of life. Finally, acting on a tip, he flies to the Himalayas and claws his way up a tall, rugged mountain. At the top, sure enough, he finds a guru who supposedly knows everything about everything.

"Tell me, O wise man," the rich guy pleads, "what is the meaning of life?"

After a moment, the pensive guru speaks: "Life, my friend, is a bowl of cherries."

The rich man feels ripped off. "What!? I came all this way for you to tell me that life is a bowl of cherries?"

Startled, the guru says: "What? You mean it isn't?"

As with our rattled guru, there's a basic insecurity that comes with being a scientist. That's because every time we scientists answer a question, we raise ten . . . a hundred . . . a thousand surprising new ones, any of which has the potential of scuttling the scientific status quo.

I love the way the famous nineteenth-century biologist and Uncertain Atheist T. H. Huxley put it: "The great tragedy of science—the slaying of a beautiful hypothesis by an ugly fact."

And Einstein too: "No amount of experimentation can ever prove me right; a single experiment can prove me wrong."

This disquieting truth explains why so many scientific ideas from even just a century ago are now obsolete. It's enough to give any conscientious scientific guru a bad case of the yips.

As a physicist and journalist, I witness this pins-and-needles uncertainty every day in ways big and small. Here's a tiny sampling of reports that crossed my desk during just a two-week period one recent summer:

- (2 August) "After 18,000 years of slimming, our planet has suddenly turned tubby round the middle. Researchers are baffled by the bulge."

- (9 August) "While global water models warn of parched days ahead, scientists worry that another pressing scarcity is information. Answers to even simple questions remain elusive: How much local water, on average, is there? How much, precisely, is going to farmers versus city dwellers? The knowledge gap shows little signs of improving."

- (12 August) "NASA's Chandra X-ray Observatory has found two stars—one too small, one too cold—that reveal cracks in our understanding of the structure of matter."

- (15 August) "Johns Hopkins scientists have revealed that so-called 'jumping genes' create dramatic re-arrangements

> in the human genome. 'These things are happening by
> mechanisms never before described,' says Jef Boeke, PhD."

Many of my scientific colleagues would have you believe this
constant process of discovery represents merely the incremental
tweaking of a logical worldview that's otherwise pretty stable.
But that, to put it mildly, is so much wishful thinking.

If the history of science has taught us anything, it's this:
seemingly insignificant little observations, or wild-eyed pro-
posals made by complete unknowns—often outsiders or outright outcasts—routinely lead to enormous upheavals in the
way science thinks about the universe.

Two of my favorite books are full of such examples, many of
them wonderfully entertaining: *They All Laughed* by Ira Flatow
and *The Experts Speak* by Christopher Cerf and Victor Navasky.

In my own book, *Five Equations That Changed the World: The
Power and Poetry of Mathematics*, I recount the particularly
poignant story of the brilliant nineteenth-century English physi-
cian and amateur scientist, Thomas Young. As a young man in his
twenties, he had the crazy notion that light consists of waves, not
particles, as everyone else in his day believed.

When the upstart Young presented his idea to the august
Royal Society in London, its members—the cream of the world
scientific community—rose up like seventeenth-century
Inquisitors and burned him at the stake with a vicious tongue-
lashing. "We now dismiss . . . the feeble locubrations of this
author," one member sneered, "in which we have searched without success for some traces of learning, acuteness, and ingenuity."

Talk about intellectual arrogance! Thomas Young, the poor man, was so devastated he temporarily quit physics.

Yet guess what? Years later, scientific experiments *vindicated* Young's wave theory. What's more, a century hence, his "feeble locubrations" played a key role in the development of quantum theory, which itself revolutionized the whole of physics!

Science, in other words, fits perfectly the definition of a chaotic system, in which tiny effects can have huge, unexpected consequences. For example:

- Board a plane at Gate 2B instead of Gate 2A, and you could find yourself winging thousands of miles away from where you wanted to be.

- Say one wrong word to someone who's on the verge of a nervous breakdown, and he or she just might shoot you.

- Let a small puff of warm, humid air waft off the southwestern African coast at just the right moment, and it could snowball across the South Atlantic and become a full-fledged hurricane. It's what meteorologists call the Butterfly Effect: one example of why weather is so fickle and unpredictable.

But wait! declares the optimist, *maybe science has experienced its last Thomas Young–like upheaval. From now on, maybe it's all about simply spit-polishing the chrome of existing theories.*

Yes, maybe. Problem is, anyone who's ever thought so has

ended up having to eat crow: for example, the famous physicist William Thomson, AKA Lord Kelvin. Talk about a towering IQ: Thomson entered the University of Glasgow at age 10, was appointed its coveted Professor of Natural Philosophy at age twenty-two, then went on to become one of the most brilliant scientists of the 1800s—and possibly all time.

As recounted in the February 2001 issue of *Natural History* magazine: at the close of the nineteenth century, the seventy-six-year-old, Irish-born mastermind looked back with extreme satisfaction at what science had accomplished. "There is nothing new to be discovered in physics now," Thomson reportedly proclaimed to a majestic gathering of the British Association for the Advancement of Science. "All that remains is more and more precise measurement [i.e., spit polishing]."

(By the way, Thomson wasn't the only scientist of his time with stars in his eyes. His younger contemporary Albert A. Michelson, the American physicist and future Nobel Prize winner, prophesied: "The more important fundamental laws and facts of physical science have all been discovered, and . . . the possibility of their being supplanted in consequence of new discoveries is exceedingly remote.")

Lord Kelvin died seven years after giving his rosy speech, which was probably for the best, because during the next two decades, a scruffy nobody named Einstein appeared out of the blue and knocked the props out from under the precious theories Kelvin and his fellow scientists had held so near and dear. Included among the casualties were nothing less than the two-hundred-year-old twin pillars of physics: Newton's laws of mechanics and his universal law of gravity.

Make no mistake: the ensuing mayhem didn't change science's view of the world slightly, or even a lot; it changed it *completely*.

Before Einstein, science was telling us we live in a three-dimensional universe in which everyone's perception of space and time is identical. After Einstein, science did a complete about-face, telling us we live in a four-dimensional universe in which everyone's perception of space and time is *not* identical.

It didn't stop there. After the recent, unexpected appearance of something called *string theory,* science appears to be in the midst of changing its mind yet again. It's now proposing we live in a universe that has *ten or more* dimensions! A decade from now, a century from now, who knows what science will be telling us about how our world really works?

The take-home lesson of all of this: worshipping the scientific process is like building your house on constantly shifting sands. As we've seen, all the evidence indicates that science is not converging smoothly and consensually upon one firm, reliable understanding of the way our world began or how it operates, really. It's not even converging upon such an understanding in fits and starts.

All the evidence indicates it's not converging at all, but rather changing its mind constantly in wildly unpredictable ways, always in concert with the latest, equally unpredictable outcomes of ever more clever and sophisticated experiments—like the wild fluctuations of the daily stock market.

Its effort to stay current and not become stale and dogmatic is science's greatest strength but also its greatest weakness. It's not a put-down of science. But for atheists who feel enormously smart about having science for their god, it should be a huge lesson in humility.

2. Logic and Reality Itself Are Insurmountably Uncertain

I find it very significant that math and science, the two greatest expressions of humanity's collective IQ, has each discovered a hard-and-fast limit to what it'll be able to teach us about the universe. In math, it's called *Gödel's theorem*. In science, it's called the *Heisenberg uncertainty principle*.

Kurt Gödel was a brilliant, twentieth-century mathematician whose specialty was logic. Like others of his day, including the famous atheist and brilliant mathematician Bertrand Russell, Gödel believed logic was the key to absolute certainty—a magic bullet with which you could rip through all the wishy-washiness that accompanies subjectivity and really *prove* things.

Gödel, however, discovered life isn't that simple.

Take a look at this logical paradox (which, if true, is false and if false, is true):

This statement is false.

According to Gödel's theorem, there will always be mathematical problems as intractable and bedeviling as that conundrum. Problems where IQ completely fails us; where the magic bullet of logic doesn't go straight to the heart of the matter but whirls around in circles, leaving our minds feeling permanently dizzy.

Gödel's scientific counterpart was Werner Heisenberg, the famous twentieth-century physicist who specialized in quantum mechanics. That's the study of how nature behaves at the tiniest, most subatomic levels. Heisenberg and others of his time believed that if we waited long enough and worked hard enough, then science would eventually reveal *everything* about the universe.

Like Gödel, however, Heisenberg had a startling revelation.

Think about how we measure, say, tire pressure. We attach one of those pen-shaped gauges to the valve and watch how far the plastic rod is spit out.

But in doing that, our gauge has had to steal away a little of the air from inside the tire, so the number we get is not the actual pressure, but the pressure inside the tire *before* we took the measurement. Who knows what the tire pressure is now, *after* the measurement?

According to Heisenberg's uncertainty principle, such tiny, residual uncertainties in our scientific study of nature are absolutely unavoidable. In the process of investigating the world, we always must sample it—catch its light, collect its sound, stroke its texture, inhale its smells, lap up its tastes—and in doing so, inevitably we alter it; not by much, but enough so that we can never be certain what the world is really like, here and now; we can only be sure of what it was like moments *before* we adulterated it with our scientific efforts to know it.

In summary, because of Gödel and Heisenberg, we now

realize science is not all-knowing and all-seeing. Unlike Dorothy in the land of Oz, our science will never be able to throw open the curtain to reveal the little lever-pulling wizard that controls everything we see.

Despite our cleverest, most assiduous mathematical and scientific efforts, the subtlest subatomic details of God's creation—and God Himself—will always remain shrouded in mystery.

3. There Will Always Be More Than One Way to Explain Things

In math, when you finish solving a problem, in many cases, you can go one step farther and prove that yours is the one and only possible solution. It's a very cool feature of mathematics called a *uniqueness proof*.

In science, however, there's no such thing, as illustrated in this terrific little story by the legendary physicist Otto Frisch, codiscoverer of nuclear fission:

"Why did Jones break his leg?"

"Because his tibia hit the curb," says the surgeon.

"Because some fool dropped a banana skin," says Mrs. Jones.

"Because he never looks where he goes," says a colleague.

"Because he subconsciously wanted a holiday," says a psychiatrist.

The point is, there's never only one way to explain things. I call it *the multiplicity principle*.

We see this principle at play in debate societies, where one person is able to argue opposing sides of an issue with equal persuasion, and among eyewitnesses to a crime, ten of whom will interpret what they saw ten different ways. And also in science, where the fertile marriage of observation and interpretation invariably spawns a multiplicity of theories. For example, consider the following real-life mystery.

Once upon a time, North America was overrun with all kinds of spectacular mammals: mammoths, saber-toothed cats, camels, bear-sized beavers, cheetahs, lions, you name it. But roughly eleven thousand years ago, something happened that killed them all off. Who or what dunnit, you ask?

Well, hold on to your hat, because scientists have come up with, not one, not two, but at least *three* equally clever, equally plausible theories:

- The big-chill theory: eleven thousand years ago, the last major ice age was coming to an end. Things were heating up too fast for the mammals to handle, so they all died out.

- The big-kill theory: shortly before eleven thousand years ago, the first humans migrated from Asia to North America across the then existing Bering land bridge. They were lean and mean and hunted the mammals to extinction.

- The big-ill theory: the first humans to cross into North America brought with them a deadly, contagious disease. That's what killed off all the mammals, not hunting.

Which do you think it was: the big chill, the big kill, or the big ill?

In the years to come, you can bet scientists will discover new information that'll help them sort out the mystery. But you can also bet the ambiguity will not come to an end.

Talk about irony. The intellectual cleverness that scientists depend on to make sense out of their observations will also always be able to conjure up more than just one explanation. For that reason, our efforts to understand the world solely with our IQ will always be doomed to permanent indecision and debate.

Our Imperfections

Human imperfection: that's the other insuperable shortcoming our IQ has in common with our SQ.

Try as we will to do the right thing—and by the way, I believe most of us do, most of the time—there's something in our nature, a stubborn flaw, that always causes us to mess up. The defect is there from birth; I see it even in my very young son. I call it the Frankenstein Effect, and here's why.

When I was a teenager, I loved watching those old, black-and-white, Universal Studios horror flicks, especially *Frankenstein*, produced in 1931 by Carl Laemmle Jr. As a budding scientist who loved toying with chemicals and electricity, I identified strongly with the single-minded Dr. Victor Frankenstein.

My favorite part of the movie—the most telling about real-life science and its effect on society—comes when Dr.

Frankenstein is sitting at a table, triumphantly puffing away at a cigar. He's just given life to the monster. Enter his elder colleague, Dr. Waldman, voice of caution and concern:

WALDMAN: This creature of yours should be kept under guard. Mark my words, he'll prove dangerous!

FRANKENSTEIN: Dangerous! Poor old Waldman. Have you never wanted to do anything that was dangerous? Where should we be if nobody had tried to find out what lies beyond? Have you never wanted to look beyond the clouds and the stars? Or what changes the darkness into light? But if you talk like that, people call you crazy. Well, if I could discover just one of these things—what eternity is, for example—I wouldn't care if they did think I was crazy!

WALDMAN: You're young, my friend—your success has intoxicated you. Wake up and look facts in the face!

The underlying message of the movie (and Mary Shelley's novel) is clear: science is a mixed blessing.

Physically, there's no question but that science has improved public health in many ways. On April 2, 1999, the U.S. Centers for Disease Control and Prevention (CDC) listed what it considered science's ten greatest achievements in this area: vaccinations, motor-vehicle safety, safer workplaces, control of infectious diseases, improved treatments for coronary heart disease and stroke, safer and healthier foods, healthier mothers and babies, family planning, fluoridation of drinking water, and recognition of tobacco as a health hazard.

The net result of these impressive achievements? "Since 1900, the average lifespan of persons in the United States has lengthened by greater than 30 years," the CDC report says, "25 of this gain attributable to advances in public health." In 1900, we lived into our forties; today we live well into our seventies and beyond.

Psychologically, however, it's a completely different story. During the past century, science has created shocking Frankensteinlike monsters that now make our lives spookier than ever. Shelley sums it up perfectly and chillingly in describing the good doctor's surprise at what he'd wrought:

> He would hope . . . that this thing which had received such imperfect animation would subside into dead matter . . . that the silence of the grave would quench forever the transient existence of the hideous corpse which he had looked upon as the cradle of life. [But] the horrid thing stands at his bedside, opening his curtains and looking on him with yellow, watery, but speculative eyes.

In his best-selling book *Why Things Bite Back: Technology and the Revenge of Unintended Consequences*, science historian Edward Tenner elaborates on the same grim truth. It's not nice to fool with Mother Nature, he warns, because: "It is the tendency of the world around us to get even, to twist our cleverness against us. Wherever we turn, we face the ironic unintended consequences of mechanical, chemical, biological, and medical ingenuity."

This Frankenstein Effect—or, as Tenner calls it, "Law of Unintended Consequences"—hit me over the head some years ago when I accepted an invitation to be the principal expert on an *Oprah Winfrey Show* devoted to everyday hazards. For an entire hour, Oprah and I listened to terrifying stories from people whose loved ones had fallen victim to everything from air bags and toy balloons to sick buildings and baby car seats to amusement-park rides and even the drawstrings on hooded sweatshirts.

It was depressing—the most dismal thing I've ever done on television—because it illustrated in gloomy detail the perilous environment our science and technology have inadvertently created for us. We now live in a man-made world full of booby traps and warning labels.

Not that life before science was any picnic—far from it. In the days of the American pioneers, men were trampled to death by runaway horses, women died while giving birth, children died while being born.

But here's the awful rub: in the process of reducing those old hazards by 90-plus percent, science has accidentally introduced new hazards that are just as lethal, if not more so:

- Who knew, for example, that our constant use of modern antibiotics would stimulate the rise of deadly, Terminator-like microbes that positively resist treatment?

- Who knew that DDT, the supposed miracle pesticide, would nearly wipe out peregrine falcons by causing the shells of their eggs to become abnormally thin and fragile?

- Who knew that Smokey the Bear's well-meaning fire prevention program would lead to cataclysmic wildfires by making our forests more vulnerable, not less?

- Who knew that better nutrition and working conditions would help us Americans live longer than ever before, but in the process also make us—as the Centers for Disease Control and Prevention so indelicately put it recently—"fatter and lazier" than ever before? Indeed, according to the CDC, "poor diet and physical inactivity" now ranks as the second highest "actual cause of death" in the U.S., right after tobacco.

And on and on and on. It's been a real kick in the shorts, this Frankenstein Effect. The cleverer we try to be, it seems, the nastier are the outcomes. It's not at all where we expected our IQs to take us.

At the beginning of the twentieth century, we were so very hopeful, so optimistic, that science would help tame Mother Nature, bend her to our will, and in the process create a utopia for us. A kind of heaven on earth.

Typical of our innocence was a huge article published in the *New York Times* on September 28, 1913. The banner headline read: "Science on Road to Revolutionize All Existence."

In the article, one Frederick Soddy, a "scientist of interna-

tional reputation," gushed forth about the wonderful consequences that surely would follow the recent discovery of
radioactivity: "Physical science . . . places no limit to the upward
path of progress, and no end to the power and ascendancy over
nature to which the race may in due course attain . . . For the
future, science promises an endless vista of new powers, new
opportunities, and new thought."

Who knew that on July 16, 1945, the great nuclear physicist
J. Robert Oppenheimer would be standing inside a reinforced
concrete bunker at Alamogordo, New Mexico, gawking through
protective glasses at the blinding white flash of the world's first
atomic bomb? No one, certainly not Mr. Soddy, had foreseen
that outcome.

Like Shelley's Dr. Frankenstein, Oppenheimer was stunned
at what he and his eminently well-intentioned colleagues had
brought into this world. The Frankenstein Effect had struck its
first major blow in the modern age.

"Now I am become Death," Oppenheimer whispered por-
tentously, quoting the Hindu god Vishnu, "destroyer of worlds."

Who knew?

Misplaced Faith

I offer up all of these examples not to put down IQ, but to
caution atheists who would turn IQ into more than what it
is—which is quite a lot, but certainly not the answer to all of
our prayers. I also want to admonish atheists who would continue slandering God because of His uncertain nature, and His

followers for their imperfections. And finally, to invite atheists who would place their trust solely in IQ to think twice before doing so.

The famous Nobel Prize–winning author Pearl S. Buck once said: "I feel no need for any other faith than my faith in human beings." Echoing similar sentiments, the Soviet cosmonaut Gherman Titov once said (right after remarking mockingly, as many of those early cosmonauts did, that he'd never once caught any glimpse of God while in outer space): "I don't believe in God. I believe in man—his strength, his possibilities, his reason."

Given the insurmountable vagaries of intellectualism and our dismal track record at trying to improve on God's creation, such brave-sounding professions of faith make me shudder. Persons who place their faith solely in human beings, science, or reason—who seek to bring about a New Enlightenment, as many of them are fond of proclaiming—do so at their own peril, and ours. Given the clear-cut lessons of history, it seems to me a very foolish and dangerous way to go.

But then again, we shouldn't be too surprised at the fact that there will always be persons who select that route. As Solomon, one of the wisest people who ever lived, warned us in Proverbs 16:25: "There is a way that seems right to a man, but in the end it leads to death."

As the father of a four-year-old boy, I constantly pray for a better world and do my best to help it come about. But I believe Utopia will come at the hands of God, not of humankind.

Why do I believe that? Because it's written in the Bible—

and all over our history. I believe it because the alternative—believing that our minds and science will someday rescue us from the mess they've helped to create in the first place—requires an untenable leap of faith that even I, a person who's able to believe in a supreme being, can't bring myself to take.

EIGHT

Can a Smart Person Believe in God?

A clash of doctrines is not a disaster—it is an opportunity.

—Alfred North Whitehead

During my graduate studies, I was awarded an AAAS (American Association for the Advancement of Science) Mass Media Science Fellowship, which enabled me to spend a summer interning at KPIX-TV, the CBS affiliate in San Francisco. While I was there, a feature reporter aired a heartwarming story that made everyone in the newsroom gather around the television set and tear up with joy. I'll never forget it.

The story was about two young men, one blind and the other paralyzed, who shared the same dream of one day whitewater rafting down a major river. But given their severe physical challenges, how could they ever hope to accomplish such a colossal feat? Halfway through the reporter's segment, we got the answer.

By using the sightless man's arm muscles and the paraplegic's 20/20 vision, the young men were able to steer their way safely through a long and deadly stretch of fast-moving water. By cooperating, by pooling their strengths, they were able to achieve the seemingly impossible.

I'm telling you this story because I believe our IQ and SQ will do the same one day. I believe they'll stop running each other down, stop jockeying for preeminence, and start cooperating in a way that will finally bring about the impossible: a fully multidimensional, stereoscopic view of God's awesome creation; a high-SQ/high-IQ perspective on the cosmos that will, above all, please our God, but also our science.

When do I believe that eye-opening miracle will happen? As I often do when contemplating such imponderables, I look to the Bible for guidance.

In 1 Corinthians 13:12, Saint Paul writes: "For now we see in a mirror, dimly, but then [one day] face to face. Now I know in part, but then I shall know just as I also am known" (NKJV).

In other words, we will suddenly see everything in full stereo—arrive at the Ultimate Collaboration—when we come face-to-face with our Creator. For the world as a whole, Paul explains two chapters later, that'll happen "in a flash, in the twinkling of an eye, at the last trumpet" (15:52).

Unfortunately for us, he doesn't give any specific day and date. In 1 Thessalonians 5:1–3, all he says is we'll know it when it happens: "Now, brothers, about times and dates we do not need to write to you, for you know very well that the day of the Lord will come like a thief in the night. While people are saying, 'Peace and safety,' destruction will come on them suddenly, as labor pains on a pregnant woman."

The prophet Isaiah is a little more explicit, predicting the last trumpet will sound—our IQ will live in peace with our SQ—when "the wolf and the lamb will feed together, and the lion will eat straw like the ox" (65:25). Yet he, too, doesn't spell out exactly when it's going to happen.

Nevertheless, I see hopeful indications we are indeed heading in the direction of an Ultimate Collaboration. What's more, it's happening at a time when both camps, our intelligence and our spirituality, are experiencing incredible boom periods.

In the intellectual realm: Because of the Internet, youngsters living in the middle of nowhere are routinely accessing places like the Musée du Louvre, British Parliament, and U.S. Library of Congress; and doctors from metropolitan cities, via long-distance medicine, are treating patients in some of the world's remotest villages. In biology, scientists are sifting through the very essence of our physical selves. And in world politics, smart weapons are changing forever what it means for countries to go to war with one another.

In the spiritual realm: Having survived—some might say, barely—a century marked by a harrowing combination of stupendous achievements and cataclysmic disasters, polls indicate most of us are frantically looking for what's missing in our otherwise comfortable existences. Suddenly facing the autumn of their lives, baby boomers are seeking to recover the spirituality they marginalized during their headlong rush to become modern and sophisticated.

For these and other reasons, spirituality has become big business in nearly every area of our lives. In organized religion, there's a building boom in nondenominational megachurches, each of which is attracting huge crowds of spiritually hungry men, women, and young people. In the workplace, luncheon prayer groups are becoming more and more popular. In the travel industry, people who wish to boost their SQs are schlepping halfway around the world just to stay at some noted spiritual retreat.

In the literary marketplace, at a time when more people are watching TV than reading books, they're flocking to Costco, Wal-Mart, and their local book dealers to snap up the latest spiritual title. Even in Hollywood—for many, the belly of the godless beast—popular celebrities such as Mandy Moore and Mel Gibson are defying Tinseltown's secular moguls by making spiritual movies that have box-office clout.

Hopeful Signs

The best news of all: even as both camps enjoy a mushrooming popularity, each is reaching out to the other in many meaningful ways. It's what encourages me to think we're heading in the right direction. Here are just three examples of what I mean.

1. Sir John Marks Templeton

I believe that through the far-flung efforts of his handsomely funded foundation, nintey-one year-old John Templeton is doing more than any single individual alive today to promote and finance the Ultimate Collaboration. Born in 1912 and reared as a Presbyterian in a small Tennessee town, Templeton attended Yale, won a Rhodes Scholarship, then worked his way to the top of the financial world by becoming what *Money* magazine described as "arguably the greatest global stock picker of the century." In 1987, Queen Elizabeth knighted him.

Today, to the tune of $40 million a year, the John Templeton Foundation supports a slew of science and religion efforts designed to increase our "knowledge and love of God." They include:

- The Spiritual Transformation Scientific Research Program, which seeks to understand what happens exactly to a person whose life is changed by religion.

- Templeton Research Lectures on the Constructive Engagement of Science and Religion, designed quite simply to get the two camps speaking to one another with open-mindedness and mutual respect.

- Campaign for Forgiveness Research and Institute for Research on Unlimited Love, designed to study two of religion's most essential themes, forgiveness and altruism.

- Science and Spirituality Centers on college campuses nationwide, designed to expose students and faculty to virtuous collaborations between IQ and SQ.

- *Science & Theology News*, a monthly international newspaper edited by Dr. Harold Koenig of Duke University.

- George Washington Institute for Spirituality and Health, located at the renowned George Washington University Medical Center. Its director, Dr. Christina Puchalski, a remarkable woman whom I featured on *Good Morning America*, is working with the Association of American Medical Colleges to introduce spirituality into medical students' training.

On top of all these efforts, the Templeton Foundation also sponsors the ultraprestigious Templeton Prize. The Prince of Wales awards this annually at Buckingham Palace to some high-IQ/high-SQ person (my characterization) whose lifelong efforts have contributed substantially to humankind's spiritual advancement.

Past winners include such religious icons as Billy Graham and Mother Teresa. But others have been persons distinguished in both camps, science and religion; for example, the Reverend Dr. John Polkinghorne, who is both a mathematical physicist and an Anglican priest. The prize is currently valued at more than a million dollars (£795,000 sterling), but by Sir John's decree it will always be worth more money than a Nobel Prize.

2. The National Institutes of Health (NIH)

The NIH is our country's premier medical research establishment, comprising 27 separate institutes located on a sprawling 322-acre campus in suburban Maryland. Its wide-ranging missions are as numerous as the entries in a medical encyclopedia, with several thousand scientists working 24/7 to find treatments and cures for everything from AIDS to macular degeneration to weapons-grade anthrax poisoning to the common cold.

Late last century, a remarkable trend in public health caused the NIH to do something extremely unorthodox. Studies showed that Americans were spending more money on so-called alternative medical treatments—an eclectic mixture of folk, Old World, anecdotal, and spiritual remedies—than on conventional, mainstream medicine. Consequently, in 1998, amid great political fanfare, the NIH created the National Center for Complementary and Alternative Medicine (NCCAM).

With an annual budget in the range of $100 million, the NCCAM is now exploring a vast universe of alleged treatments involving herbs, oils, acupuncture, homeopathy, light therapy, magnetic therapy, and the like. In the spiritual realm, the NCCAM is exploring alleged treatments such as the laying on of hands ("energy healing"), meditation, and, yes, even prayer ("remote healing").

Who knows what the NCCAM's research will teach us—its work has only just begun. What's important is that up until 1998 the scientific establishment was simply dismissing these practices out of hand, especially the spiritually related ones. Now, thanks to the NCCAM, science is at last broadening its mind—the Intellectual Cyclops is widening its view—to consider thoughtfully and respectfully beliefs held by billions of sober-minded persons worldwide. In my opinion, this is a huge step forward.

3. The American Association for the Advancement of Science (AAAS)

The AAAS is the largest general scientific society and publisher of science in the world. At 156 years old, it can be somewhat stodgy at times, but in 1995 it made a very gutsy decision. It created a new division called the Dialogue on Science, Ethics, and Religion (DoSER).

I say *gutsy* because DoSER's existence flies in the face of strongly voiced demands by leading scientists who oppose anything having to do with religion, which in their opinion is science's inferior. The distinguished seventy-six-year-old cosmologist Allan Sandage, who at fifty converted to Christianity from near atheism, explains this all too pervasive, hostile mind-set: "Today the scientific community so scorns faith, there is a reluctance to reveal yourself as a believer, the opprobrium is so severe."

Admittedly, the AAAS created DoSER with certain self-serving ulterior motives in mind. For example, it wants to reach out to religious communities nationwide, in hopes of dissuading them from supporting today's widespread efforts to have creationism taught in public school science classes.

Even so, its mission statement avers that DoSER is committed to facilitating a genuine "collaboration among scientists, ethicists, and religion scholars and leaders" to address critical social issues. To that end, it sponsors everything from large national academic conferences to small local seminars to monthly public lectures, all of which tackle meaty questions that are important to both our IQ and SQ, such as: Did the universe have a beginning? Was the universe designed? Are we alone? All in all, I believe DoSER represents a significant step in the direction toward the Ultimate Collaboration.

Caveat: not everyone is as optimistic or happy as I am about the various efforts I've just described. Intellectual Cyclopes are furious at the NIH for spending precious time and money to put spiritual claims to the test. Remember their argument? (See chapter 5.) *Everyone knows there's nothing to this spiritual stuff, so why even bother looking through the telescope?*

Spiritual Cyclopes have their own issues with the efforts I've described. They complain that Templeton's millions are luring the scientific hordes into hallowed territory, where their filthy feet are tramping all over religion's marble floors. They worry science might end up demystifying and defiling some of our most sacred spiritual mysteries.

I understand where each camp is coming from. But that's also why I'm certain the John Templeton Foundation, NIH, AAAS, and others like them are leading us in the direction of the Ultimate Collaboration: they're getting hammered equally from both sides!

Meeting of the Minds—and Souls

In addition to the foregoing efforts, there's another positive sign I find intriguing. Science and religion are becoming more and more *compatible*—or so it seems. Here are four examples of what I mean.

1. Miracles

In Genesis 21:2, we're told God caused Abraham's infertile wife, Sarah, to become pregnant, even though she was ninety years old. Intellectual Cyclopes ridicule this report as nothing but a religious fable. Sarah's pregnancy defies all the known laws of human reproduction, they claim, and besides, there's no such thing as a miracle.

But hold on. With today's rapid advances in fertility medicine, which, at the time of this printing, has already succeeded in making a sixty-three-year-old woman pregnant, the miracle is no longer so far-fetched.

As the American Society of Reproductive Medicine put it recently: "[Modern science] makes pregnancy feasible in virtually any woman with a normal uterus, regardless of age or the absence of ovaries and ovarian function."

To be sure, Intellectual Cyclopes can still decry the Bible's story about Sarah on the grounds they don't believe in miracles, but no longer on the grounds that it flies in the face of science— it doesn't. In principle, a ninety-year-old woman can indeed become pregnant. On that, the Bible and science now agree.

2. Supernatural Entities

When the thirteenth-century theologian Thomas Aquinas was first taught Aristotle's logic and science, which for generations had been lost to Western civilization, Aquinas got so excited, he spent his life trying to reconcile it with mainstream Christianity. Many in his day thought Aquinas was cuckoo because he started contemplating questions that seemed silly to them, such as: Can an angel move from A to B, without passing through all the points in between? Can two or more angels occupy the same place at the same time?

Now, in light of modern physics, Aquinas's mental exercises don't seem nearly as bizarre. Substitute for the word *angel* the phrase *subatomic particle,* and you have two questions that today's quantum theorists routinely contemplate. Namely:

- Can a subatomic particle move from A to B, without passing through all the points in between? (Answer: Yes. An electron does it constantly, in going from one atomic energy level to another—like an elevator that materializes at each floor, without passing from one to the other.)

- Can two or more subatomic particles occupy the same place at the same time? (Answer: Yes. Physicists call such entities *bosons.* Theoretically, an infinite number of them can fit on the head of a pin.)

3. Divine Creation

For centuries, religion and science have been at odds about how the universe started. According to Judaism, Christianity, and Islam, the universe had a definite beginning, created by God *ex nihilo*—out of nothing. By contrast, for a very long time, cosmologists believed in just the opposite of that, in what is called *the perfect cosmological principle*. Among other things, it holds that our universe had no definite beginning or end, but instead has existed and will exist always.

Since the mid-twentieth century, however, that scientific view has changed by 180 degrees, and in the process, come into greater agreement with the essence of Genesis 1:1 in two profound ways. First, it now appears the cosmos did indeed have a definite beginning: starting with a "big bang," it sprang into existence *out of nothing* (what scientists call *a quantum vacuum*). Second, there appears to be stunning evidence the universe came into being not accidentally, but in accordance with some master plan.

Astronomers have coined various catchy phrases to describe the million-and-one details about the cosmos that seem just too good to be random. One of them is *the rare earth hypothesis*, meaning the earth looks to be not just another planet, one among billions and billions, but a world unique in ways that really matter to us. My favorite is the *Goldilocks principle*, meaning our planet is neither too much of this nor too little of that—but just right.

For example: if the various atomic and nuclear forces affecting electrons, protons, and neutrons had been just a smidgeon weaker or stronger, the atoms we see today wouldn't exist—they'd either fly apart or collapse. Without atoms, there'd be no elements, no molecules, nothing but a chaotic soup of quantum particles.

The same goes for the force of gravity. Had it been just a tad weaker or stronger, the stars we see today wouldn't exist; they'd either blow apart or implode under their own weight. Without stars, there'd be no sun, no solar system, no earth, no imaginable life whatsoever—only cosmological bedlam.

How very, very easily our universe could've turned out disastrously! Is it merely coincidence that it didn't? That it came out looking and behaving so beautifully?

Most atheists believe so. But to many astronomers, that's way too much to swallow. The cosmos, they say, seems custom-made, fine-tuned by a Creator who knew what He was doing.

Even the late great cosmologist and Uncertain Atheist Sir Fred Hoyle, whom I had the pleasure of knowing during my years at Cornell, felt compelled to admit it. In his book *Intelligent Universe*, Hoyle says, "The probability of life originating at random is so utterly miniscule as to make it absurd."

Hoyle calculated that the chances of Randomness creating a single protein molecule by accident were about as tiny as a sightless person solving Rubik's Cube.

Put another way: he calculated Randomness would need about *three hundred times* the age of the earth to create a single protein molecule by accident. That amounts to about 1.3 trillion years! Considering that scientists currently estimate the entire universe to be only 13 *billion* years old, that doesn't leave nearly enough time for Randomness to have pieced life together from scratch, by accident.

To my knowledge, Hoyle never got around to figuring out the even greater odds against Randomness accidentally creating not just a single protein molecule, but an entire living, breathing organism. He didn't need to; he'd already made his point.

Still, off the top of my head, I'd say the chances of star dust accidentally coming together to form even a simple creature are akin to the odds of a raw chunk of marble accidentally being carved by wind and rain into Michelangelo's *Pieta*.

Concluded Hoyle before his death in 2001: "A common sense interpretation of the facts suggests that a super-intellect has monkeyed with physics, as well as with chemistry and biology . . . The numbers one calculates from the facts seem to me so overwhelming as to put this conclusion almost beyond question."

Astronomer and former borderline atheist Allan Sandage, celebrated for his exacting measurements of the size and age of the universe, put it even more bluntly: "I find it quite improbable that such order came out of chaos . . . God to me is a mystery, but is the explanation for the miracle of existence, why there is something instead of nothing."

John Eccles famous Nobel Prize—winning neurophysiologist, concurs: "There is a divine Providence over and above the materialistic happenings of biological evolution."

4. Reason Versus Faith

The Bible claims faith is precious and progressive, something that grows stronger with spiritual maturity. Hebrews 11:6 teaches that faith is as necessary to our having a meaningful relationship with God as logic is to our having one with geometry: "Without faith it is impossible to please God, because anyone who comes to him must believe that he exists and that he rewards those who earnestly seek him."

Intellectual Cyclopes, however, think of faith as being inferior to reason. For them, faith is valueless and subhuman, a shameful vestige of our benighted past.

For centuries, science appeared to be on their side. Textbooks taught that what truly separates us from other animals, what qualifies us to be at the top of the food chain, is our ability to reason.

Dolphins have language, birds make music, otters use tools, ants and bees elect leaders and operate well-run societies, beavers reengineer the landscape, butterflies navigate across vast distances, frogs have patience, ducks show love, chimps make war. But at the end of the day, science texts maintained, we alone have the ability to think logically; to put two and two together and come up with four. Hence science's name for us: *Homo sapiens*, Latin for "wise man."

But that analysis is changing radically. In recent decades, we've learned a great deal about what constitutes intelligence and how we should measure it; as a result, we've begun to theorize that other animals have far more smarts than we once thought. I did a segment on *Good Morning America* about a scientist who's even gone so far as to create an IQ test for dogs!

Now don't misunderstand me. Our incomparable intelligence still does set us apart in the world—other animals have gray matter, but not of a quality and quantity (relative to body size) as ours. It's just that we can no longer claim that having the ability to reason is what makes us *unique*.

What does? This is where the story becomes particularly intriguing.

According to the latest evidence alleged by paleoanthropologists, humans as we know them appeared suddenly and mysteriously only a very, very short while ago. (See, for example, *Becoming Human: Evolution and Human Uniqueness*, by Ian Tattersall.) By most scientific estimates, our appearance happened less than 80,000 years ago, a mere blink of the eye in geological terms.

In his book *The Third Chimpanzee*, the highly regarded UCLA physiologist Jared Diamond refers to our abrupt and enigmatic appearance as "The Great Leap Forward." Its cause? Says Diamond, "some magic twist."

In other words, scientists have yet to reach anything resembling a common explanation for how and why we happened on the scene. All they're pretty sure about is this: "That twist," says Diamond, "produced innovative, fully modern people who proceeded to spread westward from the Near East into Europe."

And therein lies the answer to our question. The Great Leap Forward (other scientists call it *The Creative Explosion*) was momentous because we humans suddenly appeared on the earth with traits never before seen in the animal kingdom— not even in any supposed protohuman species. And what exactly were these traits that make us unique? All the things that encompass what most of us would call the human soul: art, culture, and religion.

That's right. In a revelation that I believe ranks as one of the most intriguing of the twentieth century, science now seems to have evidence that what makes us positively unique is our spirituality. We *Homo sapiens* appear to be the only species on the planet with the unambiguous capacity and desire to believe in the supernatural—most notably, in an afterlife.

Nearly everywhere that paleontologists find early human remains—from western Europe to Siberia to the New World—they find evidence of our religiousness: cave paintings; carved fertility gods; dead bodies decorated with red paint, decked out in their Sunday best, and accompanied by so-called "burial goods," which include personal items such as food, weapons, and flowers.

Elephants are known to grieve for their dead and even to inter them on the spot, but contrary to legend, there are no known centralized, ceremonial elephant "graveyards." We alone appear to be creatures with a measurable SQ.

Never mind your personal views on evolution, the all-important point here is this: even though they're still leagues apart in many crucial ways, science and religion now agree on two nontrivial issues, which previously they did not: (1) religion, not reason, appears to be what makes us unique; and (2) our spirituality—our singular ability to perceive God—far from being the vestige of some benighted past, appears to be a prominent feature of a very recent *creative explosion* or *great leap forward* in the history of the world.

To be sure, the examples I've described in this chapter don't add up to anything near the Ultimate Collaboration I'm hoping and expecting to see one day, but they're also no small potatoes. To the extent that science and religion are now that much more compatible than they were half a century ago, we can take comfort in knowing *they're heading in a positive direction*.

The Ultimate Collaboration

Even though I believe science is steadily becoming more God-friendly, not less, please note—and this is very important for you to understand—I will never go so far as to claim, as some authors do, that science and religion are *converging*.

This is not a matter of semantics. The word *converging* implies that science is becoming thoroughly like religion, and religion is becoming thoroughly like science; the two are losing their individual identities and melding into a single indistinguishable entity.

(By the way, it's my impression that Cyclopes appear to be pushing for just that. I see Spiritual Cyclopes trying to turn the Bible into a science textbook and science into a God-fearing religion. Also, I see Intellectual Cyclopes trying to turn science textbooks into Bibles and religion into a science—for example, by dismissing God just because His existence can't be proven with logic, as if He were merely some high-school geometry theorem.)

As we've already discussed, my personal interpretation of the Bible, if it is to remain credible, must avoid running afoul of some irrefutable scientific achievement or discovery (see chapter 6). But my underlying faith in God is something else entirely; it does not need scientific validation, nor will it ever obtain it, given that science can never prove (or disprove) His existence (see chapter 5).

If I were suddenly stranded on a deserted island, without access to any of my sources of academic knowledge, my faith would remain intact—maybe even become stronger. *Our underlying faith in God doesn't need science, and never will, to remain alive and well.*

We've also seen the way science needs faith (see chapter 6). But it'll never have faith specifically in God. Why not? Because, once again, by design science is neutral on the subject. If it ever allows itself to be otherwise, to allow God into its strictly circumscribed descriptions of the natural world, it will cease to be science as we know it. *Our science doesn't need God, and never will, to remain alive and well.*

That's why I'm praying for not some Ultimate Convergence or Ultimate Compromise, which, by definition, can never happen—but an Ultimate Collaboration. The difference is huge.

Those two young men featured in the KPIX-TV story didn't accomplish the impossible by converging or compromising. The sightless man didn't have to become like the paralytic man, and the paralytic man didn't have to become like the sightless man. Before, during, and after their feat, they remained basically themselves.

What made their experience so very noteworthy was the respect and acceptance they showed for one another, despite being fundamentally different. Each man recognized in the other—this was crucial—a strength he needed to get the job done. They realized they needed to pool their perfectly *complementary* talents.

It's how your two eyeballs work together. They see the world differently, owing to their opposing placements on your face—

fundamentally disparate viewpoints that will never "converge" or "compromise." But within the brain, the two distinctive viewpoints collaborate and, like our two men, succeed in doing the impossible: in a process called *stereopsis*, they pool their flat, one-dimensional images to create a single, coherent, three-dimensional picture. That's how stereo vision works.

It's also how a 3-D motion picture camera works. The camera has two lenses arranged side by side, exactly like our own eyes. Because of that, the lenses record the action from slightly different angles. When the two filmic views are projected simultaneously onto a single screen, the tangled images look hopelessly indecipherable to the naked eye.

That's where those special 3-D glasses come in. Each lens sees only one of the two views; it's completely blind to the other. But within our brains, the two disparate images collaborate, and wow! Suddenly, everything on the screen looks so real and crystal clear it leaps out at you.

The Ultimate Collaboration will be like that. Our IQ and SQ, our minds and spirits, will at long last come together in a process similar to stereopsis. Think of it. On that day, for the first time in the history of our species, our two seemingly incongruent perceptions of reality will finally square up, and wow! The full, multidimensional nature of the cosmos will leap out at us.

If I'm right, then don't ever expect Spiritual or Intellectual Cyclopes to abandon their respective approaches to reality, which are indeed very different. But do expect that one day— when the wolf and the lamb feed together, and the lion eats straw like the ox—Spiritual and Intellectual Cyclopes will discover that each has a strength the other one sorely needs in its search for truth.

Glimpses of Paradise

Fortunately, we don't have to wait until only-God-knows-when to get some inkling of this glorious Ultimate Collaboration, of what it will be like when our presently divorced minds and souls are finally and fully reconciled to one another.

In the immediate aftermath of the 9/11 terrorist attacks, for example, all of us saw the telling way in which victims turned to science for the healing of their physical injuries, and to religion for the healing of their spiritual wounds.

Explained attorney Terry McGovern after losing her mother in the World Trade Center: "Trying to connect to this deeper spirituality . . . transported me to this place of hope . . . that perhaps something better does exist, and that whatever was left of my incredibly fabulous mother in that mess was not the end of her spirit" (*Faith and Doubt at Ground Zero,* PBS documentary produced by Helen Whitney).

Throughout history, many others have stolen glimpses of that same ineffable place, where our IQ and SQ collaborate wisely and peacefully:

> Science investigates; religion interprets. Science gives man knowledge which is power; religion gives man wisdom which is control. (Martin Luther King Jr.)

Both the Holy Scriptures and Nature proceed from the Divine Word. (Galileo Galilei)

I find it as difficult to understand a scientist who does not acknowledge the presence of a superior rationality behind the existence of the universe, as it is to comprehend a theologian who would deny the advances of science. (Wernher von Braun, father of the American space program)

We know the truth, not only by the reason, but also by the heart. (Blaise Pascal)

I, too, have been fortunate enough to catch glimpses of Paradise. However, it didn't happen right away.

For years, like so many intellectuals, I tried coming to God with my mind only. As I explained earlier, I wasn't ever a full-fledged Intellectual Cyclops, but at times I did come pretty close.

My SQ's slow and steady decline began when my mom died of breast cancer at the untimely age of fifty. She'd been a faithful wife and an extremely caring mother, and she loved God with all her heart. I agonized over why her life had been so cruelly shortened.

This crisis of faith happened smack in the middle of my academic studies, so it wasn't surprising that my IQ began to fill the void left behind by my dwindling SQ. In no time at all, it took control of my life.

How do I describe what happened to me? Let me put it this way. Do you remember how Aunt Polly is always incorrectly blaming poor Tom Sawyer for things his obnoxious half brother Sid actually does wrong? During my crisis, that's pretty much how it was for my faith in God: it always got blamed for whatever my mind couldn't figure out.

With only one good eye—my mind's eye—I struggled and struggled to come up with logical answers to all the usual *why* questions of life: Why am I here? Why has the basic human condition remained unchanged since the beginning of recorded history? Why does a loving God tolerate so much evil and pain in the world? Why do we humans—and we humans alone—have an inborn impulse to pray to a supernatural being? *Why? Why? Why?*

My mind couldn't answer any of these questions to its complete satisfaction, so who took the fall? My mind? You'd think so, considering it was the one coming up short. But no: Cyclopes, you'll recall, are completely blind to their own blindness.

It would never occur to our IQs to blame themselves for being unable to answer these tough spiritual questions. It's much more self-justifying to blame our SQs, which is exactly what I did. I blamed all the injustices, paradoxes, and perplexities of life on our faith in God. More to the point, I blamed God Himself and flirted with the idea of jilting Him for Randomness.

For the record: having once toyed with being an Intellectual Cyclops, I completely empathize with people who, unable to answer these tough *why* questions, simply give up and embrace atheism. But in doing so, they're missing the point and committing a huge blunder.

The clear-cut message of all that intellectual futility is that our IQ is simply not enough. Think about it: we have just enough IQ to recognize the notion of infinity and to contemplate it somewhat, but not nearly enough to fathom it, never mind God. (Mathematicians speak casually about infinity and different degrees of infinity, but neither they nor anyone else can claim to truly comprehend infinity.)

In the face of life's most enormous mysteries, we need our SQs. If we ever hope to grapple wisely with the apparent irrationalities of human existence, we need to seek God out, not run away from Him.

Please note I'm not suggesting anything here that's fundamentally different from the way we approach science and the *how* questions of life. How do Eastern Monarch butterflies know they must leave the Northeast and winter on a small number of mountaintops in the wilds of Mexico, when they've never been there before? How do our brains store memories? How does the common cold resist all of our efforts to cure it? How do nonliving molecules come together and suddenly form life? How can we hope to understand the complex interactions of the trillions of cells in our body, when our science can't even solve the famous three-body problem, which involves the interaction of just *three* particles. *How? How? How?*

Even though science can't begin to answer all of my *how* questions, I'm not about to stop believing in it. I trust science. I have faith in it.

By the same token, I'm also not about to stop believing in

God, just because my religion (and relatively feeble human brain) can't answer all of my *why* questions. I trust God. I have faith in Him . . . and in Saint Paul's promise that one day we *will* know all the answers.

My Stereoscopic Faith

To this day my IQ and SQ still vie for preeminence. But they are far more thoroughly reconciled—more optically balanced, so to speak—than they were back during my graduate-student days.

The best way I can explain this to you is in terms of this book's central metaphor. Without getting too technical, an interesting feature of a 3-D motion picture camera is this: the deeper your subject, the more disparity there is between what the two lenses see. It has to do with angles and perspective.

Point the camera at a shallow crack in the ground, and its two lenses produce only slightly different film images. But point it at the Grand Canyon and the two images are widely dissimilar—the Grand Canyon's extreme depth has the effect of magnifying differences in perspective.

The depth of human existence must be very great indeed! I say that because human reality presents itself so very, very differently to our two lenses, IQ and SQ. We're talking here about a Grand Canyon–sized disparity.

My IQ is arrogant; my SQ is humble. My IQ teaches me that

life has rhyme and reason; my SQ teaches me it has purpose and meaning. My IQ teaches me that seeing is believing; my SQ teaches me that believing is seeing. Huge differences. Yet, rather than creating contention in my life, the two together give it its awesome Grand Canyon–like depth.

Which brings me, finally, to a mystery that's dogged me all of my life: the one I explained to you in the opening chapter. Why, at so early an age, did that little boy in the barrios of East Los Angeles become gripped with the unlikely, supernatural dream of becoming, of all things, a theoretical physicist? Why didn't he opt to become a minister, just like his father, and his father's father?

After all these years of wondering about it, I believe my stereoscopic faith has finally given me the answer. You're holding it in your hands. That's right. This book could very well be the purpose of my whole existence, the chief reason I educated myself in physics, math, and astronomy, and in the process, struggled with prejudice and close-mindedness.

The book is not exactly a historic achievement—I won't get a Nobel Prize for it—but I do hope its message makes a positive difference in your life. It certainly has in mine. By opening both of my eyes wide, my stereoscopic faith has rewarded me with at least three practical insights.

One of them is that I no longer see IQ and SQ as two competing worldviews. Instead I see them as two different takes on one coherent reality. It's what I call the First Law of Stereoscopic Faith.

First Law of Stereoscopic Faith

Our IQ and SQ offer us two very different views of the same reality.

Our intelligence calls what it sees *Truth*.

Our spirituality calls what it sees *God*.

Were I an Intellectual Cyclops, I'd talk enthusiastically about neutrinos, black holes, and quantum mechanics, but disparagingly about angels, the devil, and monotheism. But as someone who sees life stereoscopically, who relies on both IQ and SQ, I see coherence where the Cyclops sees discordance. I see that:

- Angels are ghostly beings who can pass through solid objects. Neutrinos are ghostly subatomic particles that can pass through solid objects.

- The devil is an invisible cosmological predator who consumes for an eternity anyone who gets too close to him. A black hole is an invisible cosmological predator that consumes for an eternity anyone and anything that gets too close to it.

- Monotheism teaches that one and the same God can be in many different places at once. Quantum mechanics teaches that one and the same electron can be in many different places at once.

Another practical consequence of my stereoscopic faith: I now recognize that *proof* and *belief* are equally powerful ways of obtaining real, honest-to-goodness certainty. It's what I call the Second Law of Stereoscopic Faith.

Second Law of Stereoscopic Faith

God gave us—and us alone—two powerful, well-matched abilities:

A) to prove things we find hard to believe.

B) to believe in things we find hard to prove.

The three angles of any triangle—wide, narrow, or in-between—*always* add up to 180 degrees. Hard to believe, especially when you first learn it in geometry class, but my IQ has taught me how to prove it.

Meanwhile, human beings—whether black, white or in-between—*all* deserve my equal love and respect. Hard to prove, but my SQ has taught me how to believe it.

Thanks to my stereoscopic faith, I've also discovered that given the chance, IQ and SQ can actually enhance one another synergistically.

Whenever our minds study the natural world, invariably our hearts swell with a deep-seated spiritual yearning that even atheists can feel. (See chapter 2.) As Plato put it: "For everyone, as I think, must see that astronomy compels the soul to look upwards and leads from this world to another."

Similarly, whenever our spirits earnestly seek out God, invariably our minds are broadened in a way that turns lives completely around. Saint Augustine experienced that kind of powerful conversion, after which his two eyes were opened to this profound truth: "Understanding is the reward of faith. Therefore seek not to understand that thou mayest believe, but believe that thou mayest understand."

The end result of this powerful synergy between IQ and SQ is full-blown stereoscopic faith: our very own personal Ultimate Collaboration, where wisdom, love, and humility all come together in a way that makes us truly, fully human. It's what I call the Third Law of Stereoscopic faith.

Third Law of Stereoscopic Faith

Wisdom, love, and humility are the end results of stereoscopic faith—the culmination of a synergistic process by which, in becoming smarter, we become more spiritual, and in becoming more spiritual, we become smarter.

Allow me to close by giving you some examples of what wisdom, love, and humility mean to a person with stereoscopic faith.

Wisdom

I've heard wisdom defined as the ability to see life through God's supremely high-IQ/high-SQ eyes. I like that definition, but it's easier said than done. Having that kind of wisdom often requires us to defy our simpleminded logic and common sense.

Among the religious dissenters in Communist China, the story is told of a wise old farmer with nothing but a horse and son to his name. One day, the horse escapes, and all the neighbors wag their heads at the poor farmer's bad luck. The wise old man, however, quietly keeps his faith and perseveres.

A few days later, his horse returns in the company of six wild mares. This time the neighbors marvel at his *good* luck, but the wise old man quietly keeps his faith and perseveres.

Shortly afterwards, while trying to tame one of the mares, the son shatters his leg and is crippled for life. The neighbors shake their heads at this awful turn of events, but the wise old man quietly keeps his faith and perseveres.

A short while later, the brutal Warlord sweeps through the region, drafting every able-bodied man in sight. The crippled son is passed over, and once again, the neighbors celebrate the farmer's good luck. But not the wise old man; instead, he quietly keeps his faith and perseveres.

The twists and turns of life are like the scenes in an Alfred Hitchcock thriller. Unexpected developments ambush us constantly in ways that unnerve us, mentally and emotionally.

If, like the farmer's neighbors, you're an Intellectual Cyclops, it's hard to maintain an even keel. One moment you're riding high, the next moment you're feeling low.

If, like the farmer, you're someone who's been wizened by stereoscopic faith, however, then you're sustained by an inner feeling that's totally illogical. Liberated from the strict, childish conformity imposed on you by your mind and five senses, you're able to recognize that in the seeming chaos of life there is meaning and purpose.

Christians call this wise, inner counsel *the Holy Spirit*; and I believe I'm here because of it. Because of the Holy Spirit, my poor parents were able to endure on an even keel, like the Chinese farmer, the ups and downs of my unusual childhood.

As deeply religious people, my parents believed that each child has certain unique talents given to him by God, together with a sacred charge to use those talents wisely. That's why, even though Mom and Dad knew absolutely nothing about science and couldn't fathom why God had put such an odd dream into their only son's head, they encouraged me as best they could, then hung on for what became the ride of their lives and mine.

Wisdom, not logic, carried the day. And wisdom, not logic or common sense, teaches me today to cherish the tempests in my life. Awful and excruciating as these tempests can be, in surviving them, the most trying times of my existence have taught me important lessons I couldn't have possibly comprehended any other way.

Love

I'm not referring here to the frivolous and superficial imposter that passes for love in today's popular culture. Or, for that matter, the selfish counterfeit that passes for love in evolutionary biology.

In that scientific discipline, love means taking care of your own, ensuring they survive so that your gene pool can live to fight another day. It's the kind of self-centered love found in a joke credited to the seventeenth-century British satirist Samuel Butler but which perfectly expresses today's "selfish genes" view of life: "The hen," Butler quipped, "is an egg's way of producing another egg."

God commands us to love one another, even—or, rather, especially—if they're not related to us; even if we have no genetic stake in their well-being. And please note I said God *commands* us to do this.

In the Old Testament, in Leviticus 19:34, God wasn't *requesting* or expressing a *preference* as to how we should love. He was commanding us: "The alien living with you must be treated as one of your native-born. Love him as yourself, for you were aliens in Egypt. I am the LORD your God."

In the New Testament, Jesus was no less emphatic. We're to show divine love to *all* persons, He demanded, even if they're unrelated to us. Even if they're a drag on us. Even if they misuse us, as Jean Valjean did initially the loving Bishop of Digne in *Les Miserables*.

Said Jesus in Matthew 5:43–45: "You have heard that it was said, 'Love your neighbor and hate your enemy.' But I tell you: Love your enemies and pray for those who persecute you, that you may be sons of your Father in heaven."

I've always thought I understood the concept of divine love, but it wasn't until this year that it really came home to me. And when I say *came home to me,* I mean that literally. As I explained in the opening chapter, this year, my wife, Laurel, and I decided to adopt a four-year-old Hispanic-American boy.

I don't have the space here to tell you the full, amazing story, but I will say this: not in my wildest dreams did I ever think I could love a child so utterly who was not my flesh and blood. I love Laurel, so much so that I'd die for her, but for the first time in my life I believe I'm experiencing, at the very highest level, what God means by divine love—being devoted wholeheartedly to someone in whom I have absolutely no genetic stake.

An additional note: in our long, blessed journey to become parents, Laurel and I have had the pleasure of meeting many, many big-hearted people with widely different kinds and degrees of faith. Clearly, even atheists are capable of loving an adopted child like their own, and I believe God will bless them for it. But equally clear to Laurel and me is this: the love we feel for our adopted son is enhanced enormously by our devout conviction that *God* has made it all possible. It's what makes our love *divine*—the belief that this coming together of three lonely souls was not just some wonderful accident, but a match literally made in heaven.

Humility

I was named after my late paternal grandfather, Dr. Miguel Guillén (in Spanish, pronounced mee-GHEL ghee-EN)—a fact that makes me extremely happy. Grandpa was the person I most admired as a child and still admire as a grown-up. His humility and love for God made a deep and lasting impression on me.

Grandpa was born in 1896 in Los Indios, Texas. He worked from a young age and never finished high school (his doctorate was honorary). Nevertheless, he grew up to become a well-read,

Renaissance man with a sharp mind and refined tastes. He loved collecting plants, playing the violin, writing poems, and composing hymns.

Physically, he was very Anglo-looking: a tall man with movie-actor good looks, a ruddy complexion, mustache, and mesmerizing blue eyes. He always dressed in a suit and tie and wore a buff-colored, custom-made, ten-gallon Stetson. Whenever he walked into a roomful of strangers, everyone turned to stare at him.

Above all, Grandpa was the real deal: a man so truly in love with God, he veritably glowed while putting in eighteen-plus-hour days as CLADIC's president (CLADIC, you'll recall, stands for *Concilio Latino Americano de Iglesias Cristianas* and is the oldest independent Spanish-speaking Pentecostal organization in the country). Forget the Energizer Bunny! His inexhaustible devotion to God was the closest thing I've ever seen to perpetual motion.

Even in death, Grandpa was bigger than life. His funeral cortege was so long that when we in the lead limo reached the top of the Houston cemetery's hill, we couldn't see an end to the line of cars trailing behind us. That's a measure of how many persons he touched during his decades of ministry.

Years later, after a CLADIC service in which I delivered the sermon, I had the pleasure of reminiscing with some of those people. I stood in a receiving line for a full four hours, during which hundreds of *hermanos* and *hermanas* (brothers and sisters) recounted heartwarming stories concerning my grandfather—people whom he'd married, counseled, or helped in some other way.

Just as impressive to me: Grandpa's legendary stature in the CLADIC notwithstanding, he was one of the humblest persons I've ever met. As a spiritual leader every bit as venerated by his people as Martin Luther King Jr. or Billy Graham, he could've flown first class or been chauffeured around everywhere he went. Instead, he drove himself from town to town, sometimes for hundreds and hundreds of miles at a time. I'll never forget how the radiator of his car was always covered with bugs sucked up during his long journeys.

Whenever he came to East Los Angeles, since there was no room for him at our house, Grandpa could've opted to stay at a fancy hotel suite. But did he? No. Instead, he was happy to hole up in a tiny, makeshift room in the basement of our church, where he slept on a cot among the plumbing.

Looking back on Grandpa through the eyes not merely of an adoring and only grandson, but through those of a man whose television career gives him a front-row seat to every manner of human vanity and conceit, I see clearly that his humility and love for God were not just two unrelated qualities. They were, as I will explain in a moment, evidence of his profound stereoscopic faith.

So is the fact that Grandpa's favorite Bible verse was Philippians 4:8. It's chiseled in Spanish across the large stone monument CLADIC placed on his gravesite to honor his memory:

Por lo demás, hermanos, todo lo que es verdadero, todo lo honesto, todo lo justo, todo lo puro, todo lo amable, todo lo que es de buen nombre; si hay virtud alguna, si alguna alabanza, en esto pensad. ("Finally, brothers, whatever is true, whatever is noble, whatever is right, whatever is pure, whatever is lovely, whatever is admirable—if anything is excellent or praiseworthy—think about such things.")

With his astonishing, keen-eyed, stereoscopic faith, Grandpa looked at the human condition and saw past its injustice and indecency, at the universe and saw past its stars and galaxies. As a man of high IQ and high SQ, Grandpa looked past the superficial realities of our world and saw truth, nobility, rectitude, loveliness, and all of the other multidimensional facets that make up the face of God.

That's why Grandpa loved God so much. As any person with stereoscopic faith learns ultimately, Grandpa discovered that God's presence is in the best of everything around us, the way God's absence is in the worst.

That's also why Grandpa was so humble. When a person loves God with everything he has—his IQ as well as his SQ—the scale of his existence suddenly enlarges to include more and more of God. And compared to the Infinite, let's face it, it's hard for anyone to think of himself as being very big.

Final Thought

In astronomy, the science of the infinite, I was taught something fascinating that I believe summarizes perfectly the message I've tried to share with you in this book. This is what I learned: there are not one, but *two* principal sources of light that explain why the night sky is not totally pitch-black.

The first source—readily apparent to us—is the light that comes from the moon (which in reality is dimly reflected sunlight), planets, visible stars, and human activity. The second source, not at all apparent to us, is the cumulative light that rains down on earth from billions upon billions of stars that individually are too faint for us to see.

What a perfect metaphor for life!

There are many influences in our lives whose existence is as plain as the moon in the sky. I'm speaking here about all the elements of the physical realm that we readily perceive with our corporeal senses and grapple with intellectually.

But also, as we've discussed in this book, there are many other influences in life whose existence is infinitely more subtle than those physical ones, yet every bit as real and consequential to us. They comprise a mystical realm that we—and we alone—readily perceive with our intuition and grapple with spiritually.

We can't see, hear, touch, taste, or smell them, but these spiritual realities are our principal source of purpose, consolation, and deep-seated happiness. In short, like those background stars in the evening sky, they help brighten our nights.

So, then, can a smart person believe in God? Ten years ago, when my two eyes were barely beginning to open, to find their proper balance, my answer would've been just yes. But now it's so much more than that.

Now, at this exciting stage of my life, when I see truth more stereoscopically than I ever have before, my answer is: how can a smart person *not* believe in God?

NINE

What's Your SQ?

A man is infinitely more complicated than his thoughts.

—Paul Valéry

How smart are you?

Years ago, that question meant only one thing: What's your IQ? Not anymore.

Since the early 1980s, certain psychologists—most notably Howard Gardner at Harvard University—have been kicking around the notion that there's more to IQ than meets the intellectual eye. It's called *the theory of multiple intelligences*.

According to this theory, IQ still matters. It measures our logical/mathematical/linguistic abilities and is a fairly reliable predictor of how well we'll do in school and the job market. But it's not necessarily the only intelligence that counts.

For example, according to these psychologists, there's also:

- VQ (visual/spatial intelligence)
- MQ (musical intelligence)

- BQ (body intelligence)

- EQ (emotional intelligence)

And so forth. Each intelligence, in its own unique way, helps us to lead successful lives.

Spiritual Intelligence

Which brings me to my idea. Shouldn't there be an SQ, an intelligence associated with spirituality? I think so. That's why I've developed an SQ test—an IQ test for your soul, as it were.

At the end of this chapter I'll invite you to take it. But first, take a guess: How does your SQ rate? Do you have an eye for the spiritual? When you look at the world, do you see only space and time, mass and energy, logic and reason? Or do you also see connectivity and design, purpose and meaning, faith and mystery?

Take the human brain, for instance. Do you see a happy accident of molecules and chemicals? Or do you see what Candace Pert does (she's the famous high-IQ/high-SQ neurochemist who codiscovered opiate receptors): "I don't feel awe for the brain. I feel awe for God. I see in the brain all the beauty of the universe and its order—constant signs of God's presence."

By contrast, the famous and high-IQ/low-SQ psychologist Sigmund Freud wasn't able to see God *anywhere* in the universe. To him, God was nothing but a mental illness, a pathological trick of the mind. Said he: "When a man is freed of religion, he has a better chance to live a normal and wholesome life."

Unfortunately for the good doctor's place in history, hundreds of studies now show he was completely mistaken. Today, science has a whole new respect for religion.

Why the about-face? Explains Jeff Levin, social epidemiologist and author of *God, Faith, and Health*: "As those of us who have labored in this field for many years have long suspected, the relationship between religion and health, on average and at the population level, is overwhelmingly positive. Now we can say, finally, that we know this to be true."

According to science: godliness—what I'm calling SQ—is good for us, *demonstrably* good, both mentally and physically. Compared to the average population, high-SQ people appear to:

- heal faster from illness and surgery;
- recover more easily from alcohol and substance abuse;
- cope better with stress, trauma, and emotional loss;
- be less likely to suffer from depression; and
- be more likely to feel happy and optimistic.

And on and on and on. These days, hardly a week goes by without some new study crossing my desk that appears to document the very real payoffs to living the high life—the high-SQ life, that is.

Indeed, in many ways, our Spiritual Quotient is shaping up to be more important than even our Intelligence Quotient. Our devoutness to God appears to be a fairly reliable predictor of how well we'll do not just in school or the job market, but in life as a whole, which ultimately is what matters most.

In their monumental and highly acclaimed *Handbook of Religion and Health*, published by Oxford University Press, Harold G. Koenig, MD; Michael E. McCullough, PhD; and the late David B. Larson, MD, carefully reviewed no fewer than two thousand published experiments that tested the relationship between religion and everything from blood pressure, heart disease, cancer, and stroke to depression, suicide, psychotic disorders, and marital problems. The research is still in its formative stages—not all the results are clear-cut—but, says Koenig, certain overall patterns are already emerging.

They range from the small: "People who attend church service, pray individually, and read the Bible are 40% less likely to have diastolic hypertension than those who seldom participate in these religious activities."

To the large: "Religious people live longer and physically healthier lives than their nonreligious counterparts."

The largest study to date tracked the lives and deaths of 21,204 adults for a full decade. The average results? A person who attended church at least once a week lived seven years longer than someone who didn't attend at all. Among African-Americans, the disparity was even more stunning: *fourteen years!*

Explains Koenig: a high-SQ faithfulness to God appears to benefit people of all means, educational levels, and ages.

From the young: "Religious youth show significantly lower

levels of drug and alcohol abuse, premature sexual involvement, and criminal delinquency . . . They are also less likely to express suicidal thoughts or make actual attempts on their lives."

To the old: "Elderly people with deep, personal religious faith have a stronger sense of well-being and life satisfaction than their less religious peers."

Scientists are also discovering evidence of how harmful a relatively *low* SQ can be to your health. One recent study looked at individuals whose idea of being successful means having the biggest house on the block or newest luxury car. The consequences of this secular view of life are devastating, explains Ohio State University psychologist Robert Arkin: "The cycle of materialistic pursuits is disappointing and exhausting in the long run and can make people perpetually unhappy."

Millennia ago, in Ecclesiastes 5:10, Solomon made pretty much the same observation: "Whoever loves money never has money enough; whoever loves wealth is never satisfied with his income."

None of the foregoing research, I hasten to add, should be taken as proof that God exists. As George Bernard Shaw once observed, albeit cynically: "The fact that a believer is happier than a skeptic is no more to the point than the fact that a drunken man is happier than a sober one."

The mounting medical evidence in favor of spirituality proves nothing more and nothing less than the positive consequences of God's *de facto* existence—of His being a real, meaningful part of our lives. Furthermore, it illustrates two very important things.

First, science was dead wrong to treat religion as if it were an unhealthy mental illness. Clearly, it isn't.

Second, even if you still believe religion is nothing more than a *healthy* mental illness (an argument I rebut in chapter 3), then science will need to rip apart its current theories about human health. Here is evidence, not just for some ordinary placebo effect—most of the studies control for that—but astonishing health benefits that transcend the understanding of mainstream science.

Talk about irony! Science—which by design, you'll recall, excludes anything supernatural from its delimited view of the universe—appears to be on the verge of proving that religion's otherworldly God is as material to our health, if not more so, than its own worldly expertise.

The writing is now very much on the wall, says the social epidemiologist Jeff Levin. Last century, medicine went from being utterly atheistic and mechanistic, "grounded in the view that human beings are physical bodies and nothing more," to the more liberal view that we're a combination of mind and body.

Now, in what I believe is yet another dramatic step toward the Ultimate Collaboration, medicine appears to be dropping the other shoe, embracing what some are calling a *theosomatic* view. It's the view belonging to a person—or in this case, an institution—who has suddenly acquired stereoscopic faith, a view that recognizes and acknowledges our full, three-dimensional magnificence: body, mind, and spirit.

Critics whom this revolution has thrown for a loop wonder

whether the results alleging the importance of SQ will hold up. I wonder, too; we'll just have to wait and see. In my opinion, however, the most cynical of these critics are beginning to sound like those die-hards, years ago, who kept insisting the evidence linking cigarette smoking with cancer was still too inconclusive to do anything about it.

Today, an increasing number of scientists are saying: there's already more than enough evidence in the bank—it's time to act.

For the first time in its history, for example, the prestigious and powerful Association of American Medical Colleges is now training physicians to take into account the spiritual beliefs of patients under their care. According to the January, 2004 issue of *Science & Theology News*, more than 70 of the nation's 125 medical schools—including Johns Hopkins, Yale, and Stanford—are now offering the next generation of doctors courses in religion and spirituality.

Similar historic transformations are happening in the area of mental health. As I've explained, psychologists and psychiatrists have traditionally viewed religion as little more than a superstition, an irritating nuisance, or outright pathology. As recently as 1988, based on figures taken from Koenig, et. al.'s *Handbook of Religion and Health*, only about 15 percent of all U.S. psychiatric residency training programs frequently or always included religion in their curricula. Fifteen percent!— that's scandalous.

Fortunately, the Accreditation Council on Graduate Medical Education thought so too. In 1994—at long last—its Special

Requirements for Psychiatry Residency Training mandated that: (1) young psychiatrists receive religious instruction, and (2) a patient's spiritual beliefs always be noted and taken seriously.

Likewise, for the first time ever, the American Psychiatric Association's *Diagnostic and Statistical Manual of Mental Disorders (DSM-IV)*—psychiatry's massive, professional bible—now includes a category that acknowledges the importance of religion in our lives. Explains social epidemiologist Jeff Levin in his book *God, Faith, and Health*: "This development has historical importance for medicine."

In a 1995 article published in the *Journal of Nervous and Mental Disease*, the developers of the new category acknowledged with regret "psychiatry's long-standing tendency either to ignore or pathologize the religious and spiritual dimensions of human existence." (I can see Freud turning in his grave!) They expressed the hope that the *DSM-IV*'s historic gesture would ultimately "help to promote a new relationship between psychiatry and the fields of religion and spirituality."

Bringing It Home

What about you? Where do you stand in all of this? Are you what others would call *a spiritual person*? Is your SQ well-enough developed that you stand to benefit from its myriad potential mental and physical benefits?

To find out, I now invite you to take the test. Please remember: as I explained in the opening chapter, this is not a rigorous scientific examination. Rather, it's an exercise intended to make you stop for a few minutes and take stock of whether and how spirituality actually expresses itself in your day-to-day life.

The test consists of twenty multiple-choice questions, which I created based on: (1) my own lifelong association with high-SQ persons from vastly different traditional and nontraditional belief systems, and (2) my extensive reading of the published literature, particularly scores of articles by scientists presently struggling to design reliable ways of measuring a person's spirituality, which it turns out isn't easy.

Unlike the IQ tests we all took as kids, there's no time limit here. That's the beauty of spirituality: it's not temporal! Nevertheless, try going with the answer that first comes to mind—don't overthink the questions. Also, if more than one answer appeals to you, and I predict it'll happen quite often, *go with the answer you feel most passionately about.*

Also, this is not a difficult quiz—it's not meant to stump you. With those questions that are pretty easy to figure out, please try especially hard to respond honestly. For goodness' sake, don't try making your SQ come out to be better than it is really. That would defeat the whole purpose of taking the test.

Finally, when you're done, please turn to page 163 for instructions on how to calculate your SQ.

SQ Test

Okay, so now find yourself a nice quiet place and get started. Remember, be honest: answer with your heart, as well as your head.

1. If something really unlikely but good happens to me, my first reaction is to:

   a) Shake my head in wonderment over the coincidence.

   b) Try figuring out how it could've happened.

   c) Thank God.

   d) Thank Lady Luck.

   e) Try analyzing the odds of its happening.

2. When I'm walking along the beach on a warm summer evening, watching the sun set, I'm liable to be thinking about:

   a) My problems.

   b) What an incredible world we live in.

   c) How the colors of the sunset are created.

   d) What a beautiful world God created.

   e) Whether or not we're alone in the universe.

3. If I were to have an especially vivid dream, I'd probably wake up and:

   a) Try to remember everything about it.

   b) Wonder if a dead relative was trying to communicate with me.

   c) Analyze it for hidden meaning.

   d) Wonder if maybe God was trying to say something to me.

   e) Be thrilled at the vividness of it all.

4. If someone told me he'd recently felt the presence of a mysterious force more powerful than himself, I'd probably think:

   a) I've often felt that way too.

   b) He was just intoxicated or hallucinating.

   c) I've sometimes felt that way too.

   d) Oh brother, one of those people.

   e) I remember feeling that way once.

5. When things just aren't going my way, I think:

   a) Man, this is frustrating!

   b) Bad luck runs in threes, so hang in there.

   c) Why does God hate me?

   d) What am I doing wrong!?

   e) I need to pray more.

6. When I hear about a baby being killed by some psychopath, I think:

   a) What an awful world we live in.

   b) I need to pray for everyone involved, including the killer.

   c) That poor baby and her family.

   d) How could God let something like this happen?

   e) Odds are, that psychopath had abusive parents.

7. I find myself thinking about God:

a) Continually.

b) Many times a day.

c) Once a week, usually on Saturdays or Sundays.

d) Usually only during crises or special holidays.

e) Not very often.

8. Whenever I'm doing something I know is wrong, I'm liable to:

a) Scold myself.

b) Get a little thrill.

c) Hope I don't get caught.

d) Think, *Oh well, everyone does it.*

e) Ask God to forgive me.

9. If I work in the office with someone I just can't stand, I:

a) Don't try to hide it; I'm a straight shooter.

b) Smile on the outside; snarl on the inside.

c) Do my best to exercise the golden rule.

d) Try to stay away from that person as much as possible.

e) Just wait for him to slip up, so I can nail 'im.

10. The last time I attended regularly any kind of sacred house of worship was:

   a) last week; I attend every week.

   b) A month ago.

   c) A year ago.

   d) Ages ago.

   e) Never.

11. In making plans for my life, I try to:

   a) Think of how I can make the most money.

   b) Take stock of my strengths and weaknesses.

   c) Take into account what God might want from me.

   d) Think of how I can help others.

   e) Take things one day at a time.

12. When I'm in a sacred house of worship, during the quiet interludes, I tend to use the time to:

   a) Think about my problems.

   b) Enjoy the peace and quiet.

   c) Reflect on my relationship with God.

   d) Pray for myself and those in need.

   e) Revel in the beauty of the service.

13. I would say that my relationship with God is:

   a) Nonexistent.

   b) Strong.

   c) On the upswing.

   d) Hot and cold.

   e) On the skids.

14. I think prayer is:

   a) A crock, but a good backup, just in case.

   b) Overrated.

   c) A bunch of nonsense, period.

   d) The best way to connect with God.

   e) Almost always effective, one way or another.

15. When it comes to God, I:

   a) Am very confident He exists.

   b) Go back and forth about His existence.

   c) Sometimes worry that He doesn't exist.

   d) Think He's a figment of people's imaginations.

   e) Try to imagine if He's actually a He, She, or It.

16. When I look around me, I'm inclined to think I'm:

 a) More spiritual than most people.

 b) Less spiritual than most people.

 c) About as spiritual as most people.

 d) Not very spiritual.

 e) Not spiritual at all.

17. When it comes to my life right now, I feel as if:

 a) I've found meaning and purpose.

 b) I'm still looking for meaning and purpose.

 c) I'm beginning to see meaning and purpose.

 d) I'm losing sight of any meaning and purpose.

 e) I've completely given up on finding any meaning and purpose.

18. The one thing I consider most sacred, above all else, is:

 a) My relationship with God.

 b) My potential as a human being.

 c) My loved ones.

 d) Human ideals, such as freedom and justice.

 e) Nature.

19. When I see homeless people, besides wanting to help them, I:

    a) Wonder how anyone could live that way.

    b) Wonder where their families are.

    c) Grieve for what society has done to them.

    d) Grieve for them as fellow children of God.

    e) Wonder why God permits their suffering.

20. Whenever I hear someone claim science has made a discovery that disproves the existence of God:

    a) I worry but keep on believing.

    b) I begin to have serious doubts about continuing to believe.

    c) My faith in God remains steadfast.

    d) I chuckle, knowing science can never prove nor disprove God's existence.

    e) I think, *There you go, I knew it.*

TEN

Key to the SQ Test

No coward soul is mine,
No trembler in the world's storm-troubled sphere:
I see Heaven's glories shine,
And faith shines equal, arming me from fear.

—**Emily Brontë**

Congratulations on finishing the test.

There are, of course, no right or wrong answers. The scores below simply reflect the consensus view among researchers and high-SQ men and women that, at its core, spirituality consists of two things: (1) a belief in some ultimate, sacred, and divine being or reality, and (2) a lifelong search for answers in connection with that belief.

For each of your responses, give yourself the numerical score that appears next to it in the key below. For example, on the first question , if your response was C, then award yourself five points. Do this for all twenty of your responses. Your total score is your SQ.

Remember, this isn't a scientifically rigorous examination.

Rather, I pray it gives you some ballpark idea of the degree to which your eyes are open to the spiritual dimensions of human existence.

May God bless you on your spiritual journey.

Scores

1.	a) 4	b) 1	c) 5	d) 3	e) 2
2.	a) 2	b) 4	c) 1	d) 5	e) 3
3.	a) 1	b) 4	c) 2	d) 5	e) 3
4.	a) 5	b) 2	c) 4	d) 1	e) 3
5.	a) 2	b) 3	c) 4	d) 1	e) 5
6.	a) 2	b) 5	c) 3	d) 4	e) 1
7.	a) 5	b) 4	c) 3	d) 2	e) 1
8.	a) 4	b) 2	c) 3	d) 1	e) 5
9.	a) 2	b) 3	c) 5	d) 4	e) 1
10.	a) 5	b) 4	c) 3	d) 2	e) 1
11.	a) 1	b) 2	c) 5	d) 4	e) 3
12.	a) 1	b) 2	c) 5	d) 4	e) 3
13.	a) 1	b) 5	c) 4	d) 3	e) 2
14.	a) 2	b) 3	c) 1	d) 5	e) 4
15.	a) 5	b) 2	c) 3	d) 1	e) 4
16.	a) 5	b) 3	c) 4	d) 2	e) 1
17.	a) 5	b) 3	c) 4	d) 2	e) 1
18.	a) 5	b) 2	c) 1	d) 3	e) 4
19.	a) 1	b) 2	c) 3	d) 5	e) 4
20.	a) 3	b) 2	c) 4	d) 5	e) 1

Results

Find where your total score falls within the hierarchy below:

Spiritual Genius: >90
Very Spiritual: 80–90
Spiritual: 65–79
Marginally Spiritual: 50–64
Secular: <50

Acknowledgments

My heartfelt gratitude to the following sweet souls for their assistance, encouragement, and prayers: Dr. Robert H. and Mrs. Arvella Schuller, Burton Taylor, Gerald Margolis, Frank Weimann, and Carol Noel. Also: Barbara Aragon, the entire Armendariz family, the Reverend Dr. Gilberto Alvarado, Carole Cooper, Kelly Day, Kathleen Friery, Marte Guillén, Alicia Guillén, Diana Guillén, Brian Hampton, the Reverend Juan Hernandez, William Jacobs, Mort Janklow, Dr. Tim Johnson, Nancy Kay, Rebecca and Shawn Legare, Richard Leibner, Carol and Tim Milner, Don Munro, Carry Noel, Kyle Olund, Gretchen Penner, Stan Pottinger, Elizabeth Rincon, Nat Sobel, Jim Tooher, Rabbi Len Troupp, the Reverend Kenneth Valardi, and Dorothy Vincent.

A very special thank you to my dear wife and best friend, Laurel. She's always been there when I needed her, even when it wasn't convenient for her personally. After God, she is the one most responsible for the person I am today.

God bless you all.

Bibliography

Surveying the Religious Landscape: Trends in U. S. Beliefs, George Gallup Jr. and D. Michael Lindsay (Morehouse Publishing, 1999).

Objective Hope: Assessing the Effectiveness of Faith-Based Organizations: A Review of the Literature, Byron R. Johnson, Ralph Brett Tompkins, and Derek Webb (Center for Research on Religion and Urban Civil Society, University of Pennsylvania, 2002).

American Religious Identification Survey, Barry A. Kosmin, Egon Mayer, and Ariela Keysar (The Graduate Center, The City University of New York, 2001).

Traveling Mercies: Some Thoughts on Faith, Ann Lamott (Anchor Books, 2000).

A History of God: The 4,000-year Quest of Judaism, Christianity and Islam, Karen Armstrong (Ballantine Books, 1993).

Einstein and Religion, Max Jammer (Princeton University Press, 1999).

How We Believe: Science, Skepticism, and the Search for God, Michael Shermer (Henry Holt and Company, 2003).

A Devil's Chaplain, Richard Dawkins (Weidenfeld & Nicolson, 2003).

"Memes—What Are They Good For? A Critique of Memetic Approaches to Information Processing," J. W. Polichak (*Skeptic*, 6/3: 46, 1998).

The Next American Spirituality: Finding God in the Twenty-First Century, George Gallup Jr. and Timothy Jones (Cook Communications, 2000).

"Religious Revivals in Communist China," Arthur Waldron (*FPRIWire* and *Orbis*, Spring, 1998).

"Jubilee Year of Hope for Catholicism in China," James D. Whitehead and Evelyn Eaton Whitehead (*National Catholic Reporter*, September 8, 2000).

Atheism: The Case Against God, George H. Smith (Prometheus Books, 1980).

World Christian Encyclopedia (2nd Edition), David Barrett, George Kurian, and Todd Johnson (Oxford University Press, 2001).

World Christian Trends AD 30–AD 2200: Interpreting the Annual Christian Megacensus, David B. Barrett and Todd M. Johnson (William Carey Library, 2001).

The Black Book of Communism: Crimes, Terror, Repression, by Stephane Courtois, et al. (Harvard University Press, 1999).

The Theory of Everything: The Origin and Fate of the Universe, Stephen W. Hawking (New Millennium Press, 2002).

The Search for Superstrings, Symmetry, and the Theory of Everything, John Gribbin (Little, Brown and Company, 1998).

My View of the World, Erwin Schrödinger (Ox Bow Press, 1983).

World Christian Database (Center for the Study of Global Christianity, Gordon-Conwell Theological Seminary, 2004).

"When the Saints Go Marching Out: Is American Health Care Losing Its Religion?", Gloria Shur Bilchik (*Hospitals & Health Networks*, May 20, 1998, 36–40).

"Nocturnal Habits and Dark Wisdom: The American Response to Children in the Streets at Night, 1880–1930," Peter C. Baldwin (*Journal of Social History*, Spring, 2002).

Megatrends: Ten New Directions Transforming Our Lives, John Naisbitt (Warner Books, 1988).

The Encyclopedia of Ignorance, edited by Ronald Duncan and Miranda Weston-Smith (Pocket Books, 1978).

They All Laughed, Ira Flatow (HarperCollins, 1992).

The Experts Speak: The Definitive Compendium of Authoritative Misinformation, Christopher Cerf and Victor Navasky (Pantheon Books, 1984).

Five Equations That Changed the World: The Power and Poetry of Mathematics, Michael Guillen, PhD (Hyperion, 1995).

Frankenstein, Mary Shelley (Signet Classics, 1965).

Intelligent Universe, Sir Fred Hoyle (Book Sales, 1988).

Becoming Human: Evolution and Human Uniqueness, Ian Tattersall (Harvest Books, 1999).

The Third Chimpanzee: The Evolution and Future of the Human Animal, Jared Diamond (Perennial, 1992).

Handbook of Religion and Health, Harold G. Koenig, Michael E. McCullough, and David B. Larson (Oxford University Press, 2001).

God, Faith, and Health, Jeff Levin, PhD (John Wiley & Sons, 2001).

"Religious or Spiritual Problem: A Culturally Sensitive Diagnostic Category in the DSM-IV," Robert P. Turner, David Lukoff, Ruth Tiffany Barnhouse, and Francis G. Lu (*Journal of Nervous and Mental Disease*, 183: 435–44, 1995).